Mechanics and Adaptronics

Ziel der Buchreihe ist die Publikation neuer Ansätze in der Mechanik und Adaptronik. Diese Ansätze umfassen zum einen experimentelle Themen als auch solche, die numerischen Charakters sind und erstrecken sich über diverse Größenskalen. Experimentelle, zukunftsweisen Themen beinhalten Untersuchungen auf Mikro- bis hin zur Makroebene und berücksichtigen unterschiedlichste Felder, von der Mechanik über Temperatur bis hin zu unterschiedlichen chemischen Feldern. Auch auf Seiten der numerischen Simulation werden auf Grundlage der experimentellen Untersuchungen neue multiskalige Mehrfeldmodelle entwickelt, kalibriert und validiert. Eine Besonderheit stellt hier die Thematik der Adaptronik dar, die als Ziel die Integration neuer Funktionen in tragende Bauteile durch die Kombination konventioneller Konstruktionswerkstoffe mit aktiven Werkstoffsystemen verfolgt. So werden Funktionswerkstoffe betrachtet, die Energiewandler sind und die die klassischen lasttragenden und formgebenden Funktionen um sensorische und aktorische Funktionen erweitern. Aktorische und sensorische Eigenschaften ermöglichen in Kombination mit adaptiven, schnellen Reglern eine optimale Selbstanpassung an die jeweilige Einsatzumgebung. Durch die Zusammenführung von numerischem und experimentellem Know-how auf mechanische als auch adaptronischer Seite ist es möglich, Erkenntnisse aus der Natur im Sinne des bionischen Gedankens in die Technik zu übertragen und so neue wissenschaftliche Richtungen aufzuschlagen.

Matthias C. Schulz

A New Kirchhoff-Love Beam Element and its Application to Polymer Mechanics

Springer

Matthias C. Schulz
Department of Mechanical Engineering
Institute of Mechanics and Adaptronics
Braunschweig Institute of Technology
Braunschweig, Germany

ISSN 2731-6211 ISSN 2731-622X (electronic)
Mechanics and Adaptronics
ISBN 978-3-031-06339-8 ISBN 978-3-031-06340-4 (eBook)
https://doi.org/10.1007/978-3-031-06340-4

Responsible Editor: Michael Kottusch
This Springer imprint is published by the registered company Springer Nature Switzerland AG
The registered company address is: Gewerbestrasse 11, 6330 Cham, Switzerland

A New Kirchhoff-Love Beam Element and its Application to Polymer Mechanics

Von der Fakultät für Maschinenbau
der Technischen Universität Carolo-Wilhelmina zu Braunschweig

zur Erlangung der Würde

eines Doktor-Ingenieurs (Dr.-Ing.)

genehmigte Dissertation

von: Matthias Claudio Schulz
geboren in: Berlin

eingereicht am: 26.07.2021
mündliche Prüfung am: 17.11.2021

Vorsitz: Prof. Dr. habil. Rainer Krull
Gutachter: Prof. Dr.-Ing. Markus Böl
Prof. Dr.-Ing. Daniel Juhre

2022

Preface

This dissertation intends to present a finite-element framework capable of modeling the micro-mechanical behavior of single polymer chains under external load. The topic became of interest partly due to my academic background in biotechnology and partly due to the overall research orientation towards bio-mechanics of the Institute of Solid Mechanics under Prof. Markus Böl. Interestingly, despite the large body on modeling of slender bodies, at the time of the beginning of this work, there didn't seem to be suitable beam elements for my endeavor (a fact that since has changed). Therefore, a bit more than the first half of this work is concerned with the derivation and examination of a new Kirchhoff-Love beam element. Subsequently, in combination with an established finite-element alternative to molecular dynamics simulations, with a machine learning method, and with the help of many collegues, this new formulation is applied to arrive at the framework mentioned above.

Matthias Schulz

Zusammenfassung

Das Verhalten vieler Systeme wird von den mechanischen Eigenschaften stabähnlicher Bestandteile bestimmt. Beispiele reichen von Kabeln in Tragwerken, über Polysaccharide in Biofilmen auf Schiffsrümpfen, bis zu Mikrotubuli in sich teilenden Zellen. Um solche Systeme besser zu verstehen, greifen Wissenschaftler häufig auf Simulationen zurück, in denen zur Modellierung dünner Objekte, üblicherweise als Balken bezeichnet, 1D Cosserat Kontinua verwendet werden. Dies stellt einen Kompromiss zwischen Genauigkeit und Effizienz dar. In der vorliegenden Dissertation werden neuartige Finite-Elemente Formulierungen für Balken hergeleitet, auf wichtige Eigenschaften geprüft, mit existierenden Balken-Elementen verglichen und letztendlich in der Modellierung von Polymerketten verwendet, die von einem Newtonschen Fluid umgeben sind und von einer externen Kraft gestreckt werden. Basierend auf einem Gaußschen Prozess wird für solche rechenintensive Modelle ein Ersatzmodell erstellt, das im Rahmen der Bayesschen Statistik verwendet wird, um Materialparameter einer Polymerkette aus experimentellen Daten abzuleiten.

Die neuartigen Finite-Elemente Formulierungen gehören in die Kategorie der geometrisch exakten Kirchhoff-Love Balken, die sich dadurch auszeichnen, dass Schubverformung durch die Entfernung zweier Freiheitsgrade unterbunden werden. Des Weiteren weisen sie nicht nur translatorische, sondern auch rotatorische Freiheitsgrade auf und die Gesamtheit der möglichen Konfigurationen bilden einen nicht-additiven und nicht-kommutativen Raum. Deswegen werden aufwändige Interpolationsmethoden benötigt, die nicht nur wie üblicherweise auf numerisches Locking, Konvergenzverhalten und Energieerhaltung, sondern auch auf Beobachter- und Wegunabhängigkeit geprüft werden müssen. Es wird analytisch und in numerischen Beispielen überprüft, ob die drei neu vorgestellten Balken-Elemente diese Ansprüche erfüllen. Dabei werden zwei verschiedene Strategien zur Parametrisierung der Rotationen verwendet: Die aus vielen Simo-Reissner Balken gut bekannte geodätische Interpolation und die weniger bekannte Unterteilung in die sogenannte *kleinste Rotation* und einen Drehanteil. Die erstgenannte Strategie führt zu einem gemischten Balken-Element, das intrinsisch frei von numerischem Locking ist. Außerdem wird das erste Kirchhof-Love

Element präsentiert, das axiale Dehnungen durch die Entfernung eines weiteren Freiheitsgrads verhindert. Schließlich wird die numerische Effizienz der neuen Formulierungen mit verschiedenen Balken-Elementen verglichen, die Schubverformungen sowohl erlauben als auch unterbinden. Es wird geschlussfolgert, dass die neuen Elemente erhebliche numerische Vorteile für die Modellierung von dünen Balken bieten.

Molekulardynamik-Simulationen haben die Nachteile, dass sie keine ausführliche mathematische Grundlage vorweisen können und dass sie auf rechenintensive explizite Zeitintegrationsmethoden zurückgreifen müssen. Um diese Nachteile zu umgehen, ist es möglich, Polymere, die von einem Newtonschen Fluid umgeben sind, durch eine stochastische partielle Differentialgleichung zu beschreiben und mittels Finite-Elemente Anylsen Lösungen zu finden. Dabei wird das Polymer als Kirchhoff-Love Balken mit linear-elastischem Materialverhalten abgebildet, eine Flächenlast ersetzt Kollisionen des Polymers mit den Molekülen des Fluids, ein Rückwärts-Euler-Verfahren für die Zeitintegration durchgeführt und es werden sowohl Trägheitskräfte als auch elektrostatische Kräfte vernachlässigt. Die Flächenlast fluktuiert stark über Raum und Zeit und Durchschnittswerte müssen mittels der Monte-Carlo-Methode bestimmt werden. Um durch Bayesche Inferenz Unsicherheiten und Korrelationen der Materialparameter aus experimentellen Daten effizient und schnell ableiten zu können, wird ein Ersatzmodell erstellt. Hierfür wird ein Gaußscher Prozess verwendet, der für gegebene Materialparameter direkt die Durchschnittswerte des ursprünglichen Modells liefert. Die Ergebnisse werden mit der Literatur verglichen.

Summary

The physics of many systems is greatly affected by the mechanics of their slender components. Examples range from cables supporting roof structures, through polysaccharides in biofilms infesting ship hulls, to microtubules separating cells during mitosis. Aiming at a better understanding of such systems researchers often draw on simulations. In those, seeking a compromise between exactness and efficiency, it is commonly resorted to 1D Cosserat continua to model the slender components, which are commonly referred to as beams. In this thesis, novel finite element beam formulations are devised, checked for important mathematical properties, compared to existing beam formulations, and employed to simulate polymers, which are submerged in a Newtonian fluid and stretched by an external force. The emerging numerical method is used in a Monte Carlo manner to train a surrogate model based on a Gaussian process. The surrogate is needed to perform computationally efficient Bayesian inference on experimental data obtained from stretching single polymer chains.

The novel finite element formulations fall into the category of geometrically exact Kirchhoff-Love beams. A prominent characteristic of this category is that the absence of shear deformation is strongly enforced by removing two degrees of freedom. Further, the corresponding beam theories exhibit not only translational but also rotational degrees of freedom and their configurations thus form a non-additive and non-commutative space. Sophisticated interpolation schemes are required that need to be tested not only for locking, spatial convergence behavior, and energy conservation, but also for observer invariance and path-independence. For the three novel beam element formulations all these properties are analytically and numerically studied and confirmed, if applicable. Two different rotation parameterization strategies are employed based on the well-known geodesic interpolation used in many Simo-Reissner beams and the lesser known split into the so-called *smallest rotation* and a torsional part. Application of the former parameterization results in a mixed finite element formulation intrinsically free of locking phenomena. Additionally, the first geometrically exact Kirchhoff-Love beam element is presented, which strongly enforces inextensibility by removing another degree of freedom. Furthermore, the numerical efficiency of the new beam formulations is compared to other beam elements that allow for or suppress shear deformation. When modeling very slender beams, the new elements offer distinct numerical advantages.

Standard molecular dynamics simulations, which are commonly used to study polymers, suffer from a lack of a careful mathematical basis and the use of an expensive explicit time integration scheme. To circumvent these shortcomings and to be able to simulate stretching experiments on relevant time scales, the problem is described by a stochastic partial differential equation, which can be solved using the finite element method with a backward Euler temporal discretization. In detail, the polymer is represented by a Kirchhoff-Love beam with a linear elastic constitutive model. Inertial and electrostatic forces are neglected. It is deformed by a distributed load mimicking collisions with molecules of the surrounding fluid. Naturally, this load heavily fluctuates over time and space and mean values need to be computed in a Monte Carlo manner. To vastly speed up the fitting process to experimental data in a Bayesian framework, a surrogate model based on a Gaussian process is set up, which directly computes the mean values for given material parameters. The uncertainties and correlations of the material parameters are studied and compared to the literature.

Contents

List of Figures

List of Tables

Chapter 1
Introduction

A large portion of the research in the field of polymer mechanics is concerned with studying experimentally and theoretically the micro-mechanical behavior of single polymer chains and how it determines the macro-mechanical properties of the corresponding polymeric materials . Both domains are of special interest but the focus of this thesis lies on the modeling of the micro-mechanics of single polymer chains and the statistically accurate deduction of model parameters from experimental data in a Bayesian framework.

1.1 Motivation

To drastically reduce computation time when modeling the mechanical behavior of polymers it is useful to consider polymer chains as a continuum instead of a composition of single atoms. To this end, one-dimensional continuous mechanical theories are required whose governing equations depend on one variable, only. Such theories are generally referred to as beam theories (Hjelmstad, 2007), rod theories (Antman, 1995), or, when rotational degrees of freedom are used, one-dimensional Cosserat theories (Antman, 1995). Following this approach, a vast number of theoretical models have been devised, many of which were developed in the framework of statistical mechanics. Most prominently, the effort by Saitô et al. (1967) resulted in the standard model of the worm-like chain (WLC) where the polymer is considered to be an inextensible one-dimensional continuum with bending rigidity. From the mechanical point of view Fixman and Kovac (1973) provided an interesting extension to the WLC model by introducing a potential that depends on an external force stretching the chain. In parallel, new methods to conduct stretching experiments on single chains were invented, such as optical tweezers (Sun et al., 2002; Wang et al., 1997), atomic force microscopy (Bemis et al., 1999; Hugel et al., 2001; Linke et al., 2002; Zhang & Marszalek, 2006), and magnetic tweezers (Dittmore et al., 2011; Yan

M. C. Schulz, *A New Kirchhoff-Love Beam Element and its Application to Polymer Mechanics*, Mechanics and Adaptronics,
https://doi.org/10.1007/978-3-031-06340-4_1

et al., 2004). The WLC approach could be tested on different polymers by comparing theoretical and experimental results. Marko and Siggia (1995) solved the problem of a WLC under tension numerically and found good agreement to the experimental data by Smith et al. (1992), stretching single DNA molecules at high ionic strength to screen any electrostatic effects. But in the high force regime the theory failed to describe the experimental findings due to the assumption of inextensibility in the WLC model. Bouchiat et al. (1999) provided an improved fit to the numerical solution and included a correction term as proposed by Wang et al. (1997) to remedy the deficiency of a fixed contour length and to improve the overall fit to the experiments by Wang et al. (1997) on DNA molecules. Additionally this shortcoming was noticed in other stretching experiments e.g., on F-actin (Liu & Pollack, 2002), titin (Linke et al., 2002), and other polymers (Bemis et al., 1999; Hugel et al., 2001). A closed form solution for the partition function of the WLC model is not possible (Manca et al., 2012) and other physical effects such as electrostatic repulsive forces or self-contact are hard to include in the framework of statistical mechanics (Netz, 2001).

In another concept, which is followed in this work, classical mechanics is adapted to derive a kinematic description, equations of equilibrium, and a strain energy density function for the polymer chain, which is thought to be the center line of an elastic rod. A comprehensive study of this approach is given in Shi et al. (1995) and Shi et al. (1996) who, in addition to bending and axial extension, also included twist and shear deformation but unfortunately have not reported a force-extension relationship. Furthermore, in combination with such modeling approaches, in recent years the finite-element method has been applied in polymer physics with increasing frequency (Aström et al., 2008; Böl & Reese, 2005; Cyron & Wall, 2012; Cyron et al., 2013; Cyron & Wall, 2009; Heussinger & Frey, 2007; Ji & Feng, 2011; Kasas et al., 2004a; Kasas et al., 2004b; Lieleg et al., 2009; Munson et al., 2007; Schulz et al., 2019; Tserpes & Papanikos, 2005; van Dillen et al., 2006). Along the same lines, within the present work, a new take on the problem of a polymer chain under tension using a finite-element analysis is presented and validated with experimental data in a Bayesian framework. This modeling approach could be, compared to the methods of statistical mechanics, easily adapted to other physical effects such as electrostatic repulsion, self-avoidance, anisotropic chains, or even three-dimensional arrangement in a network. In addition, boundary conditions of experimental setups can be taken adequately into account. Furthermore, this numerical method exhibits reduced computational cost and a more careful mathematical basis compared to bead-spring simulations (Cyron & Wall, 2010).

1.2 Focus of This Thesis

In its first part, this thesis centers around the development of mechanical models for slender bodies that could be used to model single polymer chains and the numerical solution of the emerging partial differential equations using the finite-element method. Since shear deformation in polymers can be neglected, the focus lies on beam theories realizing such a restriction. Different kinematic descriptions of slender and initially curved bodies will be presented and used to derive strong and weak formulations of the problem. Special attention is paid to the parameterization and discretization of rotational degrees of freedom to ensure objectivity and path-independence. Further, the novel beam formulations are tested for important properties i.e., differentiability, numerical locking, conservation properties, and convergence order. Conclusively, in a thorough study of performance the presented elements are compared with results from the literature and a new beam element is identified to be used in the subsequent modeling of polymer chains.

In the second part of this thesis the beam element chosen before is used to model a polymer submerged in a Newtonian fluid. The random collisions with surrounding molecules are represented by a stochastic line load fluctuating over time and space and the emerging numerical method needs to be used in a Monte-Carlo manner to obtain mean values for given boundary conditions. It is studied how to efficiently obtain mean values. Further, to vastly accelerate the fitting process to experimental data a surrogate model is devised that directly gives the mean value in dependence of the material parameter. A Gaussian process is employed for that purpose and upon validation it is applied in a Bayesian parameter identification process based on different experimental data from the literature. This process reveals the uncertainties and correlations of the material parameters and the results are compared to the literature.

1.3 Relevance of the Present Work

Before the novel beam element can be sorted correctly into the large literature of beam elements, a short overview needs to be given tailored to the present work.

Beam theories can be categorized according to their fundamental approach. There are for example theories that result from a degeneration of the three-dimensional expressions (Bathe & Bolourchi, 1979; Dvorkin et al., 1988) or the co-rotational formulations where rigid-body rotations and local deformations are treated separately (Battini & Pacoste, 2002; Belytschko & Hsieh, 1973; Belytschko & Glaum, 1979; Crisfield, 1990; Wempner, 1969). The novel beam theory presented here is best categorized as *geometrically exact* beam element, which means that it is consistent with the virtual work principle and equilibrium equations independent of the magnitude of displacements, strains, and rotations (Crisfield & Jelenić, 1999). This term goes back to the popular Simo-Reissner beam (Reissner, 1981; Reissner, 1972; Simo, 1985), which, together with work by Cardona and Geradin (1988), Dvorkin et al. (1988), Kondoh et al. (1986), Simo and Vu-Quoc (1986), and Simo and Vu-Quoc (1988), is the basis for many subsequent studies including the present thesis.

In their initial contribution, Simo and Vu-Quoc (1986) assumed linear elasticity as well as rigid cross-sections. Stretching, shearing, bending, and twisting deformation modes were included. These basic properties can be altered and the resulting beam elements are still geometrically exact. For example, Mata et al. (2007) and Wackerfuß and Gruttmann (2009) assumed non-linear material models, Park and Lee (1996) formulated an extension to plasticity, Freund and Karako (2016) and Simo & Wong (1991) considered deformable cross-sections, and Romero et al. (2014) presented a torsion-free formulation. When shear deformation in the cross-sections is neglected, that means when the tangent to the beam axis always remains perpendicular to the cross-section, a Kirchhoff-Love beam (Kirchhoff, 1859; Love, 1944) is obtained. Such formulations were developed for initially straight configurations (Boyer & Primault, 2004; Weiss, 2002a; Weiss, 2002b; Zhang et al., 2015; Zhao & Ren, 2012), with restriction to small deformations (Armero & Valverde, 2012a; Armero & Valverde, 2012b; Greco & Cuomo, 2016; Greco & Cuomo, 2013), and in an isogeometric manner (Bauer et al., 2016). The current contribution belongs into the group of geometrically exact Kirchhoff-Love beams (Meier et al., 2014; Schulz & Böl 2019). Furthermore, it is the first contribution that is intrinsically free of locking and that can strongly enforce inextensibility.

If rotational degrees of freedom are involved, a further important property of a geometrically exact beam formulation is its rotation parameterization and interpolation strategy. In this regard, the present work employs a well-known orthogonal interpolation scheme from Crisfield and Jelenić (1999) and Jelenić and Crisfield (1999) and a new interpolation strategy similar to the one in Greco and Cuomo (2013) and Meier et al. (2014).

The application of the new beam-element in the modeling of polymers is based on the finite-element Brownian dynamics simulation (Cyron & Wall, 2009). It constitutes the first application of this method to the simulations of stretching experiments on single polymer chains. Augmented by a surrogate model the new method constitutes a new tool to study experimental data on single polymers and a flexible platform for future extensions to capture more physical effects such as contact and electrostatic repulsion.

1.4 Outline

In Sect. 2, it is focused on devising a model for slender bodies that is well suited to be used in simulations of very slender objects such as polymers. It is tried to concisely define the properties of a slender body. Two different parameterizations of rotations are presented and important non-trivial quantities like derivatives and variations are derived. Based on these preliminaries, the well-known Simo-Reissner beam is presented as reference and different Kirchhoff-Love beams are detailed, of which three are novel, and their potential advantages are discussed. They differ in the realization of the shear-stress removal as well as in the rotation parameterization and interpolation.

In Sect. 3, the presented beam theories are used to derive corresponding finite-element formulations. Special attention is paid to the temporal and spatial discretization of the orthogonal rotations. The beams are theoretically and numerically tested for objectivity, spatial convergence behavior, conservation properties, path-independence, and locking phenomena. In a thorough study of the performance the presented beam elements are tested and compared to beam elements from the literature.

In Sect. 4, the beam element, which was identified in Sect. 3 to be superior in terms of accuracy and efficiency, is employed in the finite-element Brownian dynamics simulation to model stretching experiments on single polymer chains. A fast surrogate model based on a Gaussian process is validated and applied in a Bayesian framework to obtain the uncertainties and correlations of the material parameters deducted from experimental data.

Chapter 2
Modeling of Slender Bodies

In this chapter, it is focused on devising a model for slender bodies that is well suited to be used in simulations of very slender objects such as polymers. It is tried to concisely define the properties of a slender body. Two different parameterizations of rotations are presented and important non-trivial quantities like derivatives and variations are derived. Based on these preliminaries, the well-known Simo-Reissner beam is presented to be used as as reference and different Kirchhoff-Love beams are detailed, of which three are novel, and their potential advantages are discussed. They differ in the realization of the shear-stress removal as well as in the rotation parameterization and interpolation.

2.1 A General Model for Physical Space

In this section, a concise introduction to the formal description of bodies mainly based on Chadwick (2012) and Gurtin (1982) is presented specifying terms required in the remainder of this work. For a thorough embedding of the principles of continuum mechanics in the more general theory of differential geometry the reader is referred to Aubram (2009).

A body is defined as a region $\mathcal{B}$ of the three-dimensional Euclidean space $\mathcal{E}$, which is taken here to be the collection of all 3-tupel of real numbers. $\mathcal{B}$ results from a continuously invertible map of a set of *material points* or so-called *particles* onto $\mathcal{E}$. $\mathcal{B}$ might undergo motion and deformation and is commonly referred to as *current* or *spatial* configuration. Remaining in the field of classical mechanics, these material points, which define a topological space are associated with mass and can sustain forces. Of course, other physical quantities such as thermal conductivity, entropy, or energy could be considered. Furthermore, a *reference* or *material* configuration $\mathcal{B}_0$ is defined, at least transiently. $\mathcal{B}_0$ is classically chosen to be the configuration at a

M. C. Schulz, *A New Kirchhoff-Love Beam Element and its Application to Polymer Mechanics*, Mechanics and Adaptronics,
https://doi.org/10.1007/978-3-031-06340-4_2

certain time or in a stress-free state. Also, the distinction between bodies and their closure is not made in this work. $\mathcal{B}$ signifies a connected open set joined with the respective limit points.

Each pair of points in $\mathcal{E}$ is associated with an element from the Euclidean vector space $\mathcal{V}$ over $\mathbb{R}$. Consider two arbitrary points $x, y \in \mathcal{E}$. Then, the arrow-operator

$$\rightarrow : \mathcal{E} \times \mathcal{E} \rightarrow \mathcal{V}, (x, y) \mapsto \overrightarrow{xy} \tag{2.1}$$

is introduced predicated on the axioms (Chadwick, 2012)

$$\overrightarrow{xy} = -\overrightarrow{yx} \text{ and } \overrightarrow{xz} = \overrightarrow{xy} + \overrightarrow{yz} \quad \forall x, y, z \in \mathcal{E} \tag{2.2}$$

and the parallelogram axiom that

$$\text{if } \overrightarrow{wx} = \overrightarrow{yz}, \quad \text{then } \overrightarrow{wy} = \overrightarrow{xz} \quad \forall w, x, y, z \in \mathcal{E}. \tag{2.3}$$

$\overrightarrow{xy}$ can now be visualized as an arrow pointing from x to y. Furthermore, an arbitrary point $o \in \mathcal{E}$ is chosen such that $\overrightarrow{ox}$ determines the *position* or *position vector* $\boldsymbol{x}$ of x at time t. Conversely, $\boldsymbol{x}(x, t)$ denotes the position vector of x at time t.

This association of $\mathcal{E}$ with $\mathcal{V}$ forms an affine space and enables to introduce addition, which is not meaningful in $\mathcal{E}$ but holds axiomatically in $\mathcal{V}$, to define a zero vector $\mathbf{0} \in \mathcal{V}$, which does not exist in $\mathcal{E}$, and to compute measures of distance and angle. These concepts involve the scalar and vector product on $\mathcal{V}$, which have the axiomatic properties

$$\boldsymbol{x} \cdot \boldsymbol{y} = \boldsymbol{y} \cdot \boldsymbol{x} \tag{2.4}$$

$$\boldsymbol{x} \cdot \boldsymbol{x} \geq 0 \text{ and } \boldsymbol{x} \cdot \boldsymbol{x} = 0 \text{ if and only if } \boldsymbol{x} = \mathbf{0} \tag{2.5}$$

$$(\alpha \boldsymbol{x} + \beta \boldsymbol{y}) \cdot \boldsymbol{z} = \alpha(\boldsymbol{x} \cdot \boldsymbol{z}) + \beta(\boldsymbol{y} \cdot \boldsymbol{x}) \tag{2.6}$$

and

$$\boldsymbol{x} \times \boldsymbol{y} = -\boldsymbol{y} \times \boldsymbol{x} \tag{2.7}$$

$$(\alpha \boldsymbol{x} + \beta \boldsymbol{y}) \times \boldsymbol{z} = \alpha(\boldsymbol{x} \times \boldsymbol{z}) + \beta(\boldsymbol{y} \times \boldsymbol{z}) \tag{2.8}$$

$$\boldsymbol{x} \cdot (\boldsymbol{x} \times \boldsymbol{y}) = 0 \tag{2.9}$$

for all $\boldsymbol{x}, \boldsymbol{y}, \boldsymbol{z} \in \mathcal{V}$ and $\alpha, \beta \in \mathbb{R}$, respectively. Particularly, the scalar product induces a norm i.e., a measure of length

$$||\boldsymbol{x}|| = \sqrt{\boldsymbol{x} \cdot \boldsymbol{x}} \quad \forall \boldsymbol{x} \in \mathcal{V} \tag{2.10}$$

and a measure of angle

$$\cos \phi = \frac{\boldsymbol{x} \cdot \boldsymbol{y}}{||\boldsymbol{x}|| \; ||\boldsymbol{y}||} \quad \forall \phi \in [0, \pi], \tag{2.11}$$

such that $||\boldsymbol{x}||$ is the length of $\overrightarrow{ox}$ and ϕ is the angle formed by $\overrightarrow{ox}$ and $\overrightarrow{oy}$. Furthermore, the scalar and vector product joined together satisfy the identity

$$(\boldsymbol{a} \times \boldsymbol{b}) \cdot (\boldsymbol{a} \times \boldsymbol{b}) = (\boldsymbol{a} \cdot \boldsymbol{a})(\boldsymbol{b} \cdot \boldsymbol{b}) - (\boldsymbol{a} \cdot \boldsymbol{b})^2. \tag{2.12}$$

Summarizing, we are now in a position to perform operations $(+, \cdot, \times)$ on entities of $\mathcal{V}$ and visualize them in correspondence to directed line segments in $\mathcal{E}$. For a more careful study of mathematical models for bodies the reader is referred to e.g. Gurtin (1982) and Noll (2011).

A coordinate system, sometimes also referred to as frame, (o, g) is obtained by specifying an origin $o \in \mathcal{E}$ and a basis $g = \{\boldsymbol{g}_1, \boldsymbol{g}_2, \boldsymbol{g}_3\}$ consisting of three linearly independent vectors in $\mathcal{V}$. Any position vector can then be represented as

$$\boldsymbol{x} = x_i \boldsymbol{g}_i, \tag{2.13}$$

where a repeated Latin index indicates a summation from 1 to 3. The scalars x_i denote the coordinates of $\boldsymbol{x}$ with respect to (o, g). That means, every point in $\mathcal{E}$ can be indexed by a triplet $\{x_1, x_2, x_3\} \in \mathbb{R}^3$ and the arrow-operator in Eq. (2.1) is realized by an element-wise difference of triplets. In other words, $\mathcal{E}$ is overlaid by a (e.g. cuboidal or curvilinear) number grid. Changing o translates the grid. In contrast, altering g rotates and potentially stretches this grid.

Among all possible choices for the basis vectors, the orthonormal bases $E = \{\boldsymbol{E}_1, \boldsymbol{E}_2, \boldsymbol{E}_3\}$ and $e = \{\boldsymbol{e}_1, \boldsymbol{e}_2, \boldsymbol{e}_3\}$ are introduced, which are defined by the property

$$\boldsymbol{E}_i \cdot \boldsymbol{E}_j = \boldsymbol{e}_i \cdot \boldsymbol{e}_j = \delta_{ij} \tag{2.14}$$

with the Kronecker delta

$$\delta_{ij} = \begin{cases} 1, & \text{if } i = j \\ 0, & \text{if } i \neq j. \end{cases} \tag{2.15}$$

2.2 Definition of Slender Bodies

Before continuing, it should be pointed out that there is no strict mathematical definition of slenderness of a body or for the terms *rod* and *beam*, to that effect. Nonetheless, Antman (1995) was able to specify two minimal restrictions, which are formulated in the present notation. Furthermore, in the remainder of this work the terms slender body, rod, and beam are used synonymously.

First, the concept of a *cross-section* in a body $\mathcal{B}$ is introduced. As elucidated in Sect. 2.1, every material point x can be assigned a position vector $\boldsymbol{x}$, which in turn can be indexed by a triplet $\tilde{\mathbf{x}}$ upon definition of an arbitrary coordinate system. Here, it is assumed that the three corresponding coordinates $\tilde{\mathrm{s}}_0$, $\tilde{\mathrm{x}}_2$, and $\tilde{\mathrm{x}}_3$ are obtained by continuously differentiable invertible mappings such that

$$\mathcal{B} \ni x \mapsto \tilde{\mathbf{x}}(x) \equiv (\tilde{\mathrm{s}}_0(x), \tilde{\mathrm{x}}_2(x), \tilde{\mathrm{x}}_3(x)), \tag{2.16}$$

where the applied coordinate system does not need to be specified. It is now required that the mapping $\tilde{\mathrm{s}}_0(x)$, which gives the coordinate s_0 of the material point x, ranges over the interval $(s_1, s_2) = \{s_0 : s_0 \in \tilde{\mathrm{s}}_0(\mathcal{B})\}$. Then, the cross-section $\mathcal{B}(s_0)$ is composed of all material points sharing that coordinate i.e.,

$$\mathcal{B}(s_0) := \{x \in \mathcal{B} : \tilde{\mathrm{s}}_0(x) = s_0\}. \tag{2.17}$$

Accordingly, the material points $x \in \mathcal{B}(s_0)$ are appointed the respective position vectors $\boldsymbol{x}(\bullet, \bullet, s_0, t)$.

Secondly, it is demanded that the cross-section defined in Eq. (2.17) is bounded such that any subset $\mathcal{B}(\mathcal{I})$ with $\mathcal{I}$ being a sub-interval on $[s_1, s_2]$ can be obtained by union of the according cross-sections i.e.,

$$\mathcal{B}(\mathcal{I}) := \bigcup_{s_0 \in \mathcal{I}} \equiv \{x \in \mathcal{B} : \tilde{\mathrm{s}}_0(x) = s_0, s_0 \in \mathcal{I}\}. \tag{2.18}$$

However, a body does not need to be slender in the actual sense to fulfill these two minimal restrictions. Therefore, to define the terms slenderness, rod, and beam more precisely (in the sense they are used in this work) the motion is further constrained. To this end, the center-line

$$\boldsymbol{r}(s_0, t) := \boldsymbol{x}(0, 0, s_0, t) \tag{2.19}$$

is taken to be the space curve connecting the centroids of all cross-sections. As a consequence, the first moment of area with respect to $\boldsymbol{r}(s_0, t)$ vanishes. Finally, it can be concisely defined that, in the scope of this work, the terms rod and beam denote a slender body whose position vectors are restricted to

$$\boldsymbol{x}(x, t) = \boldsymbol{r}(s_0, t) + x_2 \boldsymbol{g}_2(s_0, t) + x_3 \boldsymbol{g}_3(s_0, t), \tag{2.20}$$

where $\boldsymbol{g}_2$ and $\boldsymbol{g}_2$ are the principal axes of inertia of unit length. Appending $\boldsymbol{g}_1 = \boldsymbol{g}_2 \times \boldsymbol{g}_3$, an orthonormal, curvilinear coordinate system $(\boldsymbol{r}(s_0, t), g(s_0, t))$ naturally arises, such that x_2 and x_3 are rectangular Cartesian coordinates with regard to a coordinate system defined by its origin $r(s_0, t)$ and the basis $g(s_0, t) = \{g_1(s_0, t), g_2(s_0, t), g_2(s_0, t)\}$, which needs to be determined for every cross-section $\mathcal{B}(s_0)$. Effectively, Eq. (2.20) constraints the motion of the body and it constitutes a holonomic, rheonomic constraint, which enforces that all cross-sections always remain plane and never show warping. This is often referred to as the Bernoulli assumption.

Following the definition of a slender body presented in Eq. (2.20), an arbitrary configuration $\mathcal{B}(t)$ is uniquely defined by $\boldsymbol{r}(s_0, t)$ and the frames $\{\boldsymbol{g}_1(s_0, t), \boldsymbol{g}_2(s_0, t), \boldsymbol{g}_3(s_0, t)\}$ attached to each cross-section along the center-line. Equally, the reference configuration $\mathcal{B}_0$ is uniquely defined by $\boldsymbol{r}_0(s_0)$ and $\{\boldsymbol{g}_{01}(s_0), \boldsymbol{g}_{02}(s_0), \boldsymbol{g}_{03}(s_0)\}$. Concluding, the current and reference beam length L and L_0, respectively, are introduced such that the current curvilinear coordinate $s \in [0, L] \subset \mathbb{R}$ and the reference curvilinear coordinate $s_0 \in [0, L_0] \subset \mathbb{R}$ can be introduced.

2.3 The Special Orthogonal Group SO(3)

In the current configuration, the frame attached to each cross-section can be conveniently obtained through an orthogonal transformation via

$$\boldsymbol{g}_i(s_0) = \boldsymbol{\Lambda}(s_0)\boldsymbol{e}_i \;\text{ for }\; i = 1, 2, 3 \;\text{ with }\; \boldsymbol{\Lambda}(s_0) = \boldsymbol{g}_i(s_0) \otimes \boldsymbol{e}_i, \tag{2.21}$$

where $\otimes$ signifies the standard dyadic product. Likewise, the frame of the reference configuration can be computed as

$$\boldsymbol{g}_{0i}(s_0) = \boldsymbol{\Lambda}_0(s_0)\boldsymbol{e}_i \;\text{ for }\; i = 1, 2, 3 \;\text{ with }\; \boldsymbol{\Lambda}_0(s_0) = \boldsymbol{g}_{0i}(s_0) \otimes \boldsymbol{e}_i. \tag{2.22}$$

It is noted that these rotations are linear transformations in $\mathbb{R}^3$ and a matrix representation is available as $\boldsymbol{\Lambda} = [\boldsymbol{g}_1, \boldsymbol{g}_2, \boldsymbol{g}_3]_{\boldsymbol{E}_i}$, where the index of a matrix indicates the basis, in which the coefficients are expressed. The configuration of a rod (as defined by Eq. (2.20)) at time t can be uniquely described by a vector-valued and a tensor-valued function

$$[0, L_0] \times \mathbb{R}^+ \ni (s_0, t) \mapsto \boldsymbol{r}(s_0, t),\ \boldsymbol{\Lambda}(s_0, t) \in \mathbb{R}^3 \times SO(3). \tag{2.23}$$

$\boldsymbol{\Lambda}$ (and $\boldsymbol{\Lambda}_0$ of course) are elements of the group of proper orthogonal transformations, or *special orthogonal group* (Rossmann, 2002)

$$SO(3) := \{\boldsymbol{\Lambda} \in \mathbb{R}^{3\times 3} : \boldsymbol{\Lambda}^T = \boldsymbol{\Lambda}^{-1}, \det\boldsymbol{\Lambda} = 1\}, \tag{2.24}$$

where the standard matrix multiplication is the group operation and the identity $\boldsymbol{I}$ is the neutral element. The elements of $SO(3)$ are non-additive and non-commutative quantities thus compounding the problem of interpolating rotations in a finite-element framework, as will be addressed later. Moreover, $SO(3)$ classifies as Lie group and therefore, at every element a tangent space can be identified (Mäkinen, 2007). A schematic depiction of $SO(3)$ and other relevant quantities is shown in Fig. 2.1.

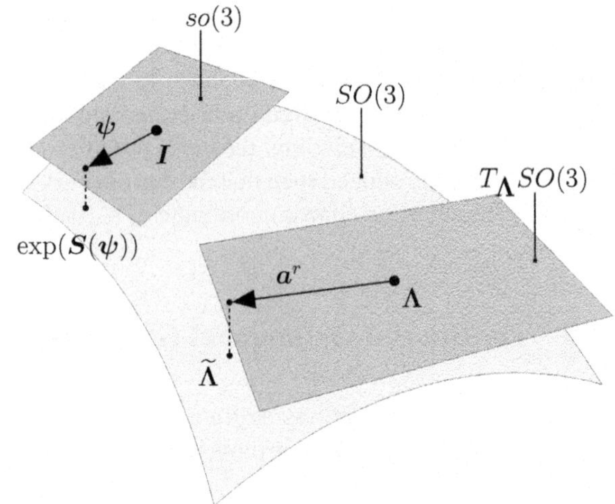

Fig. 2.1 Schematic depiction of the special orthogonal group $SO(3)$, the tangent space $so(3)$ at identity $\boldsymbol{I}$, the tangent space $T_{\boldsymbol{\Lambda}}SO(3)$ at $\boldsymbol{\Lambda}$ using a right-invariant vector field $\boldsymbol{a}^r$, and the exponential map from $so(3)$ to $SO(3)$. $\widetilde{\boldsymbol{\Lambda}}$ is obtained by transforming $\boldsymbol{a}^r$ back to $so(3)$, applying the exponential map, and transforming the obtained rotation forward

2.3.1 *The Tangent Space of SO(3) and the Corresponding Exponential Map*

It can be shown that the tangent space at the identity of $SO(3)$ consists of all skew-symmetric matrices and is represented by the set (Rossmann, 2002)

$$so(3) := \{\boldsymbol{S}(\boldsymbol{\psi}) \in \mathbb{R}^{3\times 3} : \boldsymbol{S}(\boldsymbol{\psi}) + \boldsymbol{S}(\boldsymbol{\psi})^T = \boldsymbol{0}\ \}, \tag{2.25}$$

where the operator $\boldsymbol{S}(\bullet)$ creates a skew-symmetric matrix from a vector such that $\boldsymbol{S}(\boldsymbol{a})\boldsymbol{b} = \boldsymbol{a} \times \boldsymbol{b}\ \forall \boldsymbol{a}, \boldsymbol{b} \in \mathbb{R}^3$. It is noted that $\boldsymbol{\psi}$ spanning the tangent space at the identity is an eigenvector of $\boldsymbol{\Lambda}$ (and of $\boldsymbol{\Lambda}^T$) with eigenvalue one. $so(3)$ forms the Lie algebra to $SO(3)$. It is a vector space and isomorphic to $\mathbb{R}^3$. That means every $\boldsymbol{\Lambda} \in SO(3)$ can be represented by a rotational (pseudo)-vector $\boldsymbol{\psi}$. An exponential map exists

$$so(3) \ni \boldsymbol{S}(\boldsymbol{\psi}) \mapsto \exp(\boldsymbol{S}(\boldsymbol{\psi})) := \sum_{k=0}^{\infty} \frac{1}{k!}\boldsymbol{S}(\boldsymbol{\psi})^k \in SO(3), \tag{2.26}$$

which may be evaluated efficiently by (Argyris, 1982)

$$\exp(\boldsymbol{S}(\boldsymbol{\psi})) \equiv \cos||\boldsymbol{\psi}||\boldsymbol{I} + \frac{\sin||\boldsymbol{\psi}||}{||\boldsymbol{\psi}||}\boldsymbol{S}(\boldsymbol{\psi}) + \frac{1-\cos||\boldsymbol{\psi}||}{||\boldsymbol{\psi}||^2}\boldsymbol{\psi} \otimes \boldsymbol{\psi}, \tag{2.27}$$

which in turn can be rewritten using $\boldsymbol{S}(\boldsymbol{\psi})^2 = \boldsymbol{\psi} \otimes \boldsymbol{\psi} - \boldsymbol{I}$ to (Argyris, 1982)

$$\exp(\boldsymbol{S}(\boldsymbol{\psi})) \equiv \boldsymbol{I} + \frac{\sin||\boldsymbol{\psi}||}{||\boldsymbol{\psi}||}\boldsymbol{S}(\boldsymbol{\psi}) + \frac{1-\cos||\boldsymbol{\psi}||}{||\boldsymbol{\psi}||^2}\boldsymbol{S}(\boldsymbol{\psi})^2. \tag{2.28}$$

Equations (2.27) and (2.28) are commonly attributed to Rodrigues, see e.g. Beatty (1977) for a historical review.

At other elements than the identity, there are two different ways to define a tangent space by elements of $\mathbb{R}^3$. Based on the derivatives of the left and right translation map within $SO(3)$, so-called left-invariant and right-invariant vector fields are identified, respectively, which are again isomorphic to $\mathbb{R}^3$. A left-invariant vector-field $\boldsymbol{a}^l$ spans at every $\boldsymbol{\Lambda} \in SO(3)$ the tangent space (Mäkinen, 2007)

$$T_{\Lambda}SO(3) := \{\boldsymbol{\Lambda}\boldsymbol{S}(\boldsymbol{a}^l) : \boldsymbol{a}^l \in \mathbb{R}^3\}. \tag{2.29}$$

Analogously, a right-invariant vector field $\boldsymbol{a}^r$ spans the tangent space in a different manner as (Mäkinen, 2007)

$$T_{\Lambda}SO(3) := \{\boldsymbol{S}(\boldsymbol{a}^r)\boldsymbol{\Lambda} : \boldsymbol{a}^r \in \mathbb{R}^3\}. \tag{2.30}$$

The two vector fields are related by frame transformations (Brüls et al., 2012)

$$\boldsymbol{a}^r = \boldsymbol{\Lambda}\boldsymbol{a}^l \quad \text{and} \quad \boldsymbol{S}(\boldsymbol{a}^r) = \boldsymbol{\Lambda}\boldsymbol{S}(\boldsymbol{a}^l)\boldsymbol{\Lambda}^T. \tag{2.31}$$

A schematic depiction of $SO(3)$, the tangent spaces, and the exponential map is shown in Fig. 2.1.

In the context of mechanics, left- and right-invariant vector fields can be identified as material and spatial fields (Arnold et al., 2016), respectively and thus $\boldsymbol{\Lambda}$ constitutes a push-forward operator.

It is obvious that at the identity $\boldsymbol{a}^l \equiv \boldsymbol{a}^r$. That means that, practically, a distinction between spatial and material rotation vectors $\boldsymbol{\psi}$ and $\boldsymbol{\Psi}$, respectively, could be made but that would lead to

$$\boldsymbol{\psi} = \boldsymbol{\Lambda}\boldsymbol{\Psi} = \boldsymbol{\Psi}, \tag{2.32}$$

which is not meaningful. This is also due to the fact that $\boldsymbol{\psi}$ is an eigenvector of $\boldsymbol{\Lambda}$. For more involved considerations the reader is referred e.g. to Abraham et al. (2012).

2.3.2 Derivatives and Variations of Orthogonal Transformations

The concepts of tangent spaces to $SO(3)$ and left- and right-invariant vector fields is used to implicitly define spatial and material derivatives, variations, and linearizations of $\boldsymbol{\Lambda}$. These quantities are all elements of $T_{\Lambda}SO(3)$ (Simo, 1985; Simo & Vu-Quoc, 1986). Thus, using a right-invariant representation, the spatial quantities are implicitly given by

$$\boldsymbol{\Lambda}' = \boldsymbol{S}(\boldsymbol{k})\boldsymbol{\Lambda}, \quad \delta\boldsymbol{\Lambda} = \boldsymbol{S}(\delta\boldsymbol{\theta})\boldsymbol{\Lambda}, \quad \Delta\boldsymbol{\Lambda} = \boldsymbol{S}(\Delta\boldsymbol{\theta})\boldsymbol{\Lambda}, \quad \text{and} \quad \dot{\boldsymbol{\Lambda}} = \boldsymbol{S}(\boldsymbol{w})\boldsymbol{\Lambda}, \tag{2.33}$$

where $\boldsymbol{k}$ is the spatial curvature vector, $\delta\boldsymbol{\theta}$ denotes the spatial rotation vector variation, $\Delta\boldsymbol{\theta}$ depicts the spatial multiplicative rotation vector increment, and $\boldsymbol{w}$ signifies the spatial angular velocity vector. Based on left-invariant vector fields, the material counterparts are obtained through

$$\boldsymbol{\Lambda}' = \boldsymbol{\Lambda}\boldsymbol{S}(\boldsymbol{K}), \quad \delta\boldsymbol{\Lambda} = \boldsymbol{\Lambda}\boldsymbol{S}(\delta\boldsymbol{\Theta}), \quad \Delta\boldsymbol{\Lambda} = \boldsymbol{\Lambda}\boldsymbol{S}(\Delta\boldsymbol{\Theta}), \quad \text{and } \dot{\boldsymbol{\Lambda}} = \boldsymbol{\Lambda}\boldsymbol{S}(\boldsymbol{W}). \tag{2.34}$$

Herein, $\boldsymbol{K}$ depicts the material curvature vector, $\delta\boldsymbol{\Theta}$ is the material rotation vector variation, $\Delta\boldsymbol{\Theta}$ defines the material multiplicative rotation vector increment, and $\boldsymbol{W}$ is the material angular velocity vector. The spatial and material quantities are all related by the frame transformation defined in Eq. (2.31).

Additional insight into the physical meaning of $\boldsymbol{K}$ and $\boldsymbol{k}$ can be gained by realizing that (Kapania & Li, 2003)

$$\boldsymbol{K} = [K_1, K_2, K_3]^T_{\boldsymbol{e}_i} \text{ and } \boldsymbol{k} = [K_1, K_2, K_3]^T_{\boldsymbol{g}_i}, \tag{2.35}$$

where the index of the column vector containing the coordinates denotes the basis, in which the corresponding vector is represented. Physically, K_1 is the twist around $\boldsymbol{g}_1$, and K_2 and K_3 are the curvature components around $\boldsymbol{g}_2$ and $\boldsymbol{g}_3$, respectively. The definition of the curvature introduced in this approach differs from the common definition. This difference is due to the fact that the angle increment is related to the undeformed arc-length increment $\mathrm{d}s_0$ instead of the arc-length increment of the deformed configuration $\mathrm{d}s$. The relation $\mathrm{d}s = ||\boldsymbol{r}'||\mathrm{d}s_0$ can be used to compute the mathematical definition of curvature via

$$\boldsymbol{K}_m = \frac{\boldsymbol{K}}{||\boldsymbol{r}'||} \text{ and } \boldsymbol{k}_m = \frac{\boldsymbol{k}}{||\boldsymbol{r}'||}, \tag{2.36}$$

respectively. Holzapfel & Ogden (2011) introduced the designation 'stretch-modified curvature' for $\boldsymbol{K}$ and $\boldsymbol{k}$ and argued that their use is more convenient because they are independent of axial strain, respectively.

It is instructive to further notice that the equivalence of components of spatial and material counterparts as seen in Eq. (2.35) also holds for other kinematic quantities such that

$$\delta\boldsymbol{\Theta} = [\delta\Theta_1, \delta\Theta_2, \delta\Theta_3]^T_{\boldsymbol{e}_i} \text{ and } \delta\boldsymbol{\theta} = [\delta\Theta_1, \delta\Theta_2, \delta\Theta_3]^T_{\boldsymbol{g}_i} \tag{2.37}$$

and

$$\boldsymbol{W} = [W_1, W_2, W_3]^T_{\boldsymbol{e}_i} \text{ and } \boldsymbol{w} = [W_1, W_2, W_3]^T_{\boldsymbol{g}_i}. \tag{2.38}$$

The physical interpretation of the coordinates $\delta\Theta_1$, $\delta\Theta_2$, and $\delta\Theta_3$ as well as of W_1, W_2, and W_3 is analogous to the interpretation of K_1, K_2, and K_3 given above.

2.3.3 *Parameterization of Orthogonal Rotations*

Generally speaking, a parameterization of a rotation is a mapping from an open set in the Euclidean space to another open set in $SO(3)$ and since $SO(3)$ is a compact manifold, there exists no singularity-free mapping (Stuelpnagel, 1964). This is, of course, also true for the two rotation parameterization schemes presented in the following. To circumvent this problem, Mäkinen (2007) performed a change of parameterization for rotations bigger than half a revelation. But since a reference rotation tensor will be employed within each element (see Sect. 3.2.3) , the parameterizations are locally free of singularities with the restriction that an element performs less than a rotation of π within a single load increment or time step, respectively, which is a reasonable limitation in simulations.

2.3.3.1 Parameterization Using Rotation Vectors

A commonly employed parameterization of rotations is based on the concept of the rotation vector $\boldsymbol{\psi} \in so(3)$, which was introduced in Sect. 2.3.1. The orthogonal rotation is then obtained using Rodrigues' formula from Eqs. (2.27) and (2.28) as $\boldsymbol{\Lambda} = \exp(\boldsymbol{S}(\boldsymbol{\psi}))$. Conversely, $\boldsymbol{\psi}$ is extracted from $\boldsymbol{\Lambda}$ using Spurrier's algorithm (Klumpp, 1976; Spurrier, 1978), which will be abstractly referred to as

$$\boldsymbol{\psi} = \mathrm{sp}(\boldsymbol{\Lambda}). \tag{2.39}$$

A unique extraction is only possible within the interval $||\boldsymbol{\psi}|| \in]-\pi, \pi]$.
It has become obvious that due to the orthogonality restrictions formulated in Eq. (2.24), out of the nine components of $\boldsymbol{\Lambda}$ only three are independent. That means following the parameterization strategy presented in this section, rotations are fully characterized using only three degrees of freedom.

In the finite-element framework, which is addressed in Sect. 3, increments are computed and the rotations need to be updated. Therefore, before continuing, the difference between additive and multiplicative rotation increments $\Delta\boldsymbol{\psi}$ and $\Delta\boldsymbol{\theta}$, respectively, should be clarified. Let the superscript k refer to a step within the iterative Newton-Raphson solution scheme, then

$$\boldsymbol{\Lambda}^{k+1} = \exp(\boldsymbol{S}(\boldsymbol{\psi}^{k+1})) = \exp(\boldsymbol{S}(\boldsymbol{\psi}^{k} + \Delta\boldsymbol{\psi}^{k})) = \exp(\Delta\boldsymbol{\theta}^{k}) \exp(\boldsymbol{S}(\boldsymbol{\psi}^{k})) \tag{2.40}$$

holds, but

$$\boldsymbol{\Lambda}^{k+1} \neq \exp(\boldsymbol{S}(\boldsymbol{\psi}^{k} + \Delta\boldsymbol{\theta}^{k})). \tag{2.41}$$

Having stressed this distinction, $\Delta\boldsymbol{\psi}$ and $\Delta\boldsymbol{\theta}$ are related by the tangent operator $\boldsymbol{T}$ via

$$\Delta\boldsymbol{\psi} = \boldsymbol{T}\Delta\boldsymbol{\theta}, \quad \boldsymbol{T}(\boldsymbol{\psi}) = \frac{1}{\psi^2}\boldsymbol{S}(\boldsymbol{\psi})\boldsymbol{S}(\boldsymbol{\psi})^T + \frac{\psi/2}{\tan(\psi/2)}(\boldsymbol{I} - \frac{1}{\psi^2}\boldsymbol{S}(\boldsymbol{\psi})\boldsymbol{S}(\boldsymbol{\psi})^T) - \frac{1}{2}\boldsymbol{S}(\boldsymbol{\psi}) \tag{2.42}$$

and

$$\Delta\boldsymbol{\theta} = \boldsymbol{T}^{-1}\Delta\boldsymbol{\psi}, \quad \boldsymbol{T}^{-1}(\boldsymbol{\psi}) = \frac{1}{\psi^2}(1 - \frac{\sin(\psi)}{\psi})\boldsymbol{\psi} \otimes \boldsymbol{\psi} + \frac{\sin(\psi)}{\psi}\boldsymbol{I} + \frac{1-\cos(\psi)}{\psi^2}\boldsymbol{S}(\boldsymbol{\psi}), \tag{2.43}$$

respectively, see e.g. Jelenić & Crisfield (1999). The variations $\delta\boldsymbol{\psi}$ and $\delta\boldsymbol{\theta}$ are equally related. The material curvature vector and its linearization for this parameterization strategy follow as (Crisfield & Jelenić, 1999)

$$\boldsymbol{K} = \boldsymbol{T}^{-1}(\boldsymbol{\psi})\boldsymbol{\psi}' \quad \text{and} \quad \Delta\boldsymbol{K} = \boldsymbol{\Lambda}^T\Delta\boldsymbol{\theta}'. \tag{2.44}$$

Interestingly enough, the angular velocity vector shows a similar structure when expressed in terms of the rotation vector and it follows

$$\boldsymbol{w} = \boldsymbol{T}^{-1}(\boldsymbol{\psi})\dot{\boldsymbol{\psi}} \quad \text{and} \quad \boldsymbol{a} = \dot{\boldsymbol{T}}^{-1}\dot{\boldsymbol{\psi}} + \boldsymbol{T}^{-1}\ddot{\boldsymbol{\psi}} \tag{2.45}$$

with the transformation rule $\boldsymbol{w} = \boldsymbol{\Lambda}\boldsymbol{W}$.

It has become clear that the rotation vector parameterization as proposed here exhibits a singularity when the rotation angle equals 2π or a multiple of that. The singularity is due to the compactness of $SO(3)$, which is why there does not exist a single continuous map from $\mathbb{R}^3$. For more details on that matter the interested reader is referred to Stuelpnagel (1964).

2.3.3.2 Parameterization Using a Split into *Smallest-Rotation* and Torsional Rotation

In this approach, as done by Greco & Cuomo (2013), the orthogonal transformation is represented as

$$\boldsymbol{\Lambda} = \boldsymbol{R}(\boldsymbol{g}_1, \varphi)\boldsymbol{\Lambda}_{sr}(\boldsymbol{e}_1, \boldsymbol{g}_1), \tag{2.46}$$

where

$$\boldsymbol{g}_1 = \frac{\boldsymbol{r}'}{||\boldsymbol{r}'||} \tag{2.47}$$

denotes the normalized tangent to the center-line and

$$\boldsymbol{R}(\varphi, \boldsymbol{g}_1) = \exp(\boldsymbol{S}(\varphi\boldsymbol{g}_1)) \tag{2.48}$$

represents the rotation about $\boldsymbol{g}_1$ by the torsion angle φ. $\boldsymbol{\Lambda}_{sr}$ describes the rotation about the vector (Crisfield, 1997)

$$\boldsymbol{\alpha} = \alpha\frac{\boldsymbol{e}_1 \times \boldsymbol{g}_1}{||\boldsymbol{e}_1 \times \boldsymbol{g}_1||} \quad \text{with} \quad \cos\alpha = \boldsymbol{e}_1^T\boldsymbol{g}_1 \quad \text{and} \quad \sin\alpha = ||\boldsymbol{e}_1 \times \boldsymbol{g}_1||, \tag{2.49}$$

such that based on Eq. (2.27)

$$\boldsymbol{\Lambda}_{sr}(\boldsymbol{e}_1, \boldsymbol{g}_1) = \boldsymbol{I} + \boldsymbol{S}(\boldsymbol{e}_1 \times \boldsymbol{g}_1) + \frac{1}{1 + \boldsymbol{e}_1^T \boldsymbol{g}_1} \boldsymbol{S}(\boldsymbol{e}_1 \times \boldsymbol{g}_1)^2 \tag{2.50}$$

holds. Obviously, a singularity arises for the case that $\boldsymbol{e}_1 = -\boldsymbol{g}_1$. This problem is alleviated by introducing a reference rotation tensor $\boldsymbol{\Lambda}_r$, which is chosen in each element at the beginning of every increment. This idea seems to originate in the work by Crisfield & Jelenić (1999) and Jelenić & Crisfield (1999) and was first applied by Meier et al. (2014) in combination with the present parameterization technique. Following this approach, the orthogonal transformation is given as

$$\boldsymbol{\Lambda} = \boldsymbol{R}(\boldsymbol{g}_1, \varphi)\boldsymbol{\Lambda}_{sr}(\boldsymbol{g}_0, \boldsymbol{g}_1)\boldsymbol{\Lambda}_r \quad \text{with} \quad \boldsymbol{g}_0 = \boldsymbol{\Lambda}_r \boldsymbol{e}_1. \tag{2.51}$$

Considering the tangent to the deformed center-line $\boldsymbol{t} \equiv \boldsymbol{r}'$ as a variable, it can be seen that this rotation parameterization method is realized by four degrees of freedom $(\boldsymbol{t}, \varphi)$. On the one hand, in total, this is one degree more than is needed in the parameterization using rotation vectors. On the other hand $||\boldsymbol{t}||$ does not alter the rotation, thereby actually reducing the required degrees of freedom. However, as will be seen in Sect. 3.4.2, when employing this rotation parameterization technique in Kirchhoff-Love beams, the total number of required parameters is not increased.

The Unit Sphere S^2 and Singularity: In a short excursion based on Simo & Fox (1989), Simo et al. (1990) and Romero et al. (2014), the smallest-rotation transformation is introduced from a more mathematical viewpoint starting at the unit sphere

$$S^2 := \left\{\boldsymbol{g}_1 \in \mathbb{R}^3 \mid ||\boldsymbol{g}_1|| = 1\right\}, \tag{2.52}$$

which is only a differentiable manifold but not a group like $SO(3)$ since no suitable binary operation is introduced. The according tangent space at $\boldsymbol{g}_1 \in S^2$, which is a linear space, reads

$$T_{\boldsymbol{g}_1} S^2 := \left\{\delta \boldsymbol{g}_1 \in \mathbb{R}^3 \mid \boldsymbol{g}_1 \cdot \delta \boldsymbol{g}_1 = 0\right\} \tag{2.53}$$

and the map back to S^2 follows as

$$T_{\boldsymbol{g}_1} S^2 \ni \delta \boldsymbol{g}_1 \mapsto \exp_t(\delta \boldsymbol{g}_1) := \cos ||\delta \boldsymbol{g}_1|| \boldsymbol{g}_1 + \frac{\sin ||\delta \boldsymbol{g}_1||}{||\delta \boldsymbol{g}_1||} \boldsymbol{g}_1 \in S^2. \tag{2.54}$$

Next, a straight line in the tangent space in the direction $\delta \boldsymbol{g}_1$ running through $\boldsymbol{g}_{01}$ is considered. Mapping this line onto S^2 introduces the concept of a geodesic i.e., the equivalence to a straight line in S^2. Assuming that the geodesic runs through $\boldsymbol{g}_{01}$ and $\boldsymbol{g}_1$, this connection can be considered the shortest path between these two elements. The orthogonal transformation relating $\boldsymbol{g}_{01}$ and $\boldsymbol{g}_1$ was introduced above as smallest-rotation $\boldsymbol{\Lambda}_{sr}(\boldsymbol{g}_{01}, \boldsymbol{g}_1)$. Furthermore, the relation

$$\boldsymbol{g}_{01} \cdot \boldsymbol{g}_1 = \cos(||\delta \boldsymbol{g}_1||) \quad \text{with} \quad \delta \boldsymbol{g}_1 \in T_{\boldsymbol{g}_1} S^2 \tag{2.55}$$

holds.

It has become evident that $\boldsymbol{\Lambda}_{sr}(\boldsymbol{g}_{01}, \boldsymbol{g}_1)$ exhibits a singularity for $\boldsymbol{g}_{01} = \boldsymbol{g}_1$, which could easily occur in simulations. But if a new reference tangent vector $\boldsymbol{g}_{01}$ is defined within each element at the beginning of an iteration, within a time step the maximum rotation relative to $\boldsymbol{g}_{01}$ is π, which is a reasonable restriction.

Curvature Vector: The material curvature vector in dependence of $\boldsymbol{t}$, $\boldsymbol{t}_0$, and φ reads

$$S(\boldsymbol{K}) = \boldsymbol{\Lambda}^T\boldsymbol{\Lambda}' = \boldsymbol{\Lambda}^T(\boldsymbol{R}'\boldsymbol{\Lambda}_{sr}\boldsymbol{\Lambda}_r + \boldsymbol{R}\boldsymbol{\Lambda}'_{sr}\boldsymbol{\Lambda}_r), \tag{2.56}$$

where it is stressed that $\boldsymbol{t}_0' = 0$. Then, using the relation

$$\boldsymbol{g}_1' = \frac{\boldsymbol{t}'}{||\boldsymbol{t}||} - \frac{\boldsymbol{t}^T\boldsymbol{t}'}{||\boldsymbol{t}||^3}\boldsymbol{t} \tag{2.57}$$

the spatial derivations of Eqs. (2.48) and (2.50) yield

$$\begin{aligned} \boldsymbol{R}' =& \varphi'\left(\cos\varphi \boldsymbol{S}(\boldsymbol{g}_1) + \sin\varphi \boldsymbol{S}(\boldsymbol{g}_1)^2\right) + \sin\varphi \boldsymbol{S}(\boldsymbol{g}_1') \\ &+ (1-\cos\varphi)\left(\boldsymbol{S}(\boldsymbol{g}_1)\boldsymbol{S}(\boldsymbol{g}_1') + \boldsymbol{S}(\boldsymbol{g}_1')\boldsymbol{S}(\boldsymbol{g}_1)\right) \end{aligned} \tag{2.58}$$

and

$$\begin{aligned} \boldsymbol{\Lambda}'_{sr} =& \boldsymbol{S}(\boldsymbol{t}_0 \times \boldsymbol{g}_1') + \frac{1}{1+\boldsymbol{t}_0^T\boldsymbol{g}_1}\left(\boldsymbol{S}(\boldsymbol{t}_0 \times \boldsymbol{g}_1')\boldsymbol{S}(\boldsymbol{t}_0 \times \boldsymbol{g}_1) + \boldsymbol{S}(\boldsymbol{t}_0 \times \boldsymbol{g}_1)\boldsymbol{S}(\boldsymbol{t}_0 \times \boldsymbol{g}_1')\right) \\ &- \frac{\boldsymbol{t}_0^T\boldsymbol{g}_1'}{(1+\boldsymbol{t}_0^T\boldsymbol{g}_1)^2}\boldsymbol{S}(\boldsymbol{t}_0 \times \boldsymbol{g}_1)^2, \end{aligned} \tag{2.59}$$

respectively. Alternatively, $\boldsymbol{k}$ can be split into a part parallel ($\boldsymbol{k}_{\parallel}$) and a part perpendicular ($\boldsymbol{k}_{\perp}$) to $\boldsymbol{g}_1$ such that (Meier et al., 2014)

$$\boldsymbol{k} = \boldsymbol{k}_{\parallel} + \boldsymbol{k}_{\perp} = (K_{sr} + \varphi')\boldsymbol{g}_1 + \frac{\boldsymbol{S}(\boldsymbol{t})\boldsymbol{t}'}{||\boldsymbol{t}||^2}, \tag{2.60}$$

where

$$\begin{aligned} K_{sr} &= (\boldsymbol{\Lambda}_{sr}(\boldsymbol{g}_0, \boldsymbol{g}_1)\boldsymbol{\Lambda}_r\boldsymbol{e}_2)^T(\boldsymbol{\Lambda}_{sr}(\boldsymbol{g}_0, \boldsymbol{g}_1)\boldsymbol{\Lambda}_r\boldsymbol{e}_3)' \\ &= (\boldsymbol{\Lambda}_{sr}(\boldsymbol{g}_0, \boldsymbol{g}_1)\boldsymbol{\Lambda}_r\boldsymbol{e}_3)^T(\boldsymbol{\Lambda}_{sr}(\boldsymbol{g}_0, \boldsymbol{g}_1)\boldsymbol{\Lambda}_r\boldsymbol{e}_2)' \end{aligned} \tag{2.61}$$

is the curvature evoked by the smallest-rotation part.

Rotation Vector Variation: In the following, the (additive) variations $(\delta\boldsymbol{t}, \delta\varphi)$ of $(\boldsymbol{t}, \varphi)$ are derived as they are required in the following. To this end, $\delta\boldsymbol{\theta}$ is split into different contributions as

$$\delta\boldsymbol{\theta} = \delta\boldsymbol{\theta}_{\delta\varphi} + \delta\boldsymbol{\theta}_{\delta t} + \delta\boldsymbol{\theta}_{\delta t_0}. \tag{2.62}$$

Herein, $\delta\boldsymbol{\theta}_{\delta\varphi}$, $\delta\boldsymbol{\theta}_{\delta t}$, and $\delta\boldsymbol{\theta}_{\delta t_0}$depict variations of $\delta\boldsymbol{\theta}$ due to variations in φ, $\boldsymbol{t}$, and $\boldsymbol{t}_0$, respectively, and their analytical expressions are determined via an approach, which is similar to the one used by Ibrahimbegović (1995) who transformed $\delta\boldsymbol{\psi}$ to $\delta\boldsymbol{\Theta}$.

To begin with, the relations

$$\frac{\mathrm{d}}{\mathrm{d}\,\varepsilon}\exp(\boldsymbol{S}(\varepsilon\delta\boldsymbol{\theta}))\;|_{\varepsilon=0} = \boldsymbol{S}(\delta\boldsymbol{\theta}) \tag{2.63}$$

and

$$\delta\boldsymbol{g}_1 = \frac{1}{||\boldsymbol{t}||}[\boldsymbol{I} - \boldsymbol{g}_1 \otimes \boldsymbol{g}_1]\delta\boldsymbol{t} \;\text{ and }\; \delta\boldsymbol{g}_0 = \frac{1}{||\boldsymbol{t}_0||}[\boldsymbol{I} - \boldsymbol{g}_0 \otimes \boldsymbol{g}_0]\delta\boldsymbol{t}_0 \tag{2.64}$$

and the identities (Simo & Vu-Quoc, 1986)

$$\boldsymbol{S}(\boldsymbol{a})^3 = -||\boldsymbol{a}||^2\boldsymbol{S}(\boldsymbol{a}) \;\text{ and }\; \boldsymbol{S}(\boldsymbol{a})\boldsymbol{S}(\boldsymbol{b})\boldsymbol{S}(\boldsymbol{a}) = -(\boldsymbol{a}\cdot\boldsymbol{b})\boldsymbol{S}(\boldsymbol{a}) \quad \forall \boldsymbol{a},\boldsymbol{b} \in \mathbb{R}^3 \tag{2.65}$$

are repeated here. Then, to compute $\delta\boldsymbol{\theta}_{\delta\varphi}$

$$\frac{\mathrm{d}}{\mathrm{d}\,\varepsilon}\exp(\boldsymbol{S}(\varepsilon\delta\boldsymbol{\theta}_{\delta\varphi}))\boldsymbol{R}\boldsymbol{\Lambda}_{sr}\boldsymbol{\Lambda}_r\;|_{\varepsilon=0} \equiv \frac{\mathrm{d}}{\mathrm{d}\,\varepsilon}\left[\boldsymbol{R}(\varphi+\varepsilon\delta\varphi,\boldsymbol{g}_1)\boldsymbol{\Lambda}_{sr}(\boldsymbol{g}_0,\boldsymbol{g}_1)\boldsymbol{\Lambda}_r\right]\;|_{\varepsilon=0} \tag{2.66}$$

is rearranged to

$$\frac{\mathrm{d}}{\mathrm{d}\,\varepsilon}\left[\boldsymbol{R}(\varphi+\varepsilon\delta\varphi,\boldsymbol{g}_1)\right]\;|_{\varepsilon=0}\,\boldsymbol{\Lambda}_{sr}\boldsymbol{\Lambda}_{sr}^{-1}\boldsymbol{R}^{-1} = \boldsymbol{S}(\boldsymbol{g}_1)\delta\varphi. \tag{2.67}$$

Analogously, to determine an expression for $\delta\boldsymbol{\theta}_{\delta t}$

$$\frac{\mathrm{d}}{\mathrm{d}\,\varepsilon}\exp(\boldsymbol{S}(\varepsilon\delta\boldsymbol{\theta}_{\delta t}))\boldsymbol{R}\boldsymbol{\Lambda}_{sr}\boldsymbol{\Lambda}_r\;|_{\varepsilon=0} \equiv \frac{\mathrm{d}}{\mathrm{d}\,\varepsilon}\left[\boldsymbol{R}(\varphi,\boldsymbol{g}_1+\varepsilon\delta\boldsymbol{g}_1)\boldsymbol{\Lambda}_{sr}(\boldsymbol{g}_0,\boldsymbol{g}_1+\varepsilon\delta\boldsymbol{g}_1)\boldsymbol{\Lambda}_r\right]\;|_{\varepsilon=0}\,, \tag{2.68}$$

can be solved using Eqs. (2.63), (2.64), and (2.65). Subsequently, since $\delta\boldsymbol{g}_1^T\boldsymbol{g}_1 = 0$, a lengthy but otherwise straight-forward computation results in

$$\begin{aligned}
\boldsymbol{S}(\delta\boldsymbol{\theta}_{\delta t}) &= \frac{\mathrm{d}}{\mathrm{d}\,\varepsilon}\left[\boldsymbol{R}(\varphi,\boldsymbol{g}_1+\varepsilon\delta\boldsymbol{g}_1)\boldsymbol{\Lambda}_{sr}(\boldsymbol{g}_0,\boldsymbol{g}_1+\varepsilon\delta\boldsymbol{g}_1)\right]\;|_{\varepsilon=0}\,\boldsymbol{\Lambda}_{sr}^{-1}\boldsymbol{R}^{-1} \\
&= \frac{\mathrm{d}}{\mathrm{d}\,\varepsilon}\left[\boldsymbol{\Lambda}_{sr}(\boldsymbol{g}_0,\boldsymbol{g}_1+\varepsilon\delta\boldsymbol{g}_1)\right]\;|_{\varepsilon=0}\,\boldsymbol{\Lambda}_{sr}^{-1} \\
&= \frac{1}{||\boldsymbol{t}||}\left(\boldsymbol{S}(\boldsymbol{g}_1\times\delta\boldsymbol{t}) + \boldsymbol{S}(\boldsymbol{g}_1\times\boldsymbol{g}_0)\boldsymbol{S}(\boldsymbol{g}_1\times\delta\boldsymbol{t}) - \boldsymbol{S}(\boldsymbol{g}_1\times\delta\boldsymbol{t})\boldsymbol{S}(\boldsymbol{g}_1\times\boldsymbol{g}_0)\right).
\end{aligned} \tag{2.69}$$

Alternatively, a different approach could be taken by using Eq. (2.43) in combination with a relation between $\delta\boldsymbol{\psi}$ and $(\delta\boldsymbol{t}, \delta\varphi)$ to obtain relation (2.69).

Similarly, to obtain $\delta\boldsymbol{\theta}_{\delta t_0}$ in terms of $\delta\boldsymbol{t}_0$

$$\frac{\mathrm{d}}{\mathrm{d}\,\varepsilon}\exp(\boldsymbol{S}(\varepsilon\delta\boldsymbol{\theta}_{\delta t_0}))\boldsymbol{\Lambda}_{sr}\boldsymbol{\Lambda}_r\mid_{\varepsilon=0} \equiv \frac{\mathrm{d}}{\mathrm{d}\,\varepsilon}\left[\boldsymbol{\Lambda}_{sr}(\boldsymbol{g}_0+\varepsilon\delta\boldsymbol{g}_0, \boldsymbol{g}_1)\boldsymbol{\Lambda}_r\right]\mid_{\varepsilon=0}, \tag{2.70}$$

is solved, which is a lengthy but otherwise straight-forward task where Eqs. (2.63), (2.64), and (2.65) are used. Ultimately, this computation results in

$$\begin{aligned}\boldsymbol{S}(\delta\boldsymbol{\theta}_{\delta t_0}) &= \frac{\mathrm{d}}{\mathrm{d}\,\varepsilon}\left[\boldsymbol{\Lambda}_{sr}(\boldsymbol{g}_0+\varepsilon\delta\boldsymbol{g}_0, \boldsymbol{g}_1)\right]\mid_{\varepsilon=0}\boldsymbol{\Lambda}_{sr}^{-1}\\ &= \boldsymbol{S}(\delta\boldsymbol{g}_0\times\boldsymbol{g}_1)+\frac{1}{1+\boldsymbol{g}_0^T\boldsymbol{g}_1}(-\delta\boldsymbol{g}_0^T\boldsymbol{g}_1\boldsymbol{S}(\boldsymbol{g}_0\times\boldsymbol{g}_1)\\ &\quad+\boldsymbol{S}(\boldsymbol{g}_0\times\boldsymbol{g}_1)\boldsymbol{S}(\delta\boldsymbol{g}_0\times\boldsymbol{g}_1)-\boldsymbol{S}(\delta\boldsymbol{g}_0\times\boldsymbol{g}_1)\boldsymbol{S}(\boldsymbol{g}_0\times\boldsymbol{g}_1))\end{aligned} \tag{2.71}$$

due to $\delta\boldsymbol{g}_0^T\boldsymbol{g}_0 = 0$.

Applying the definition of the commutator (Lie bracket) of skew-symmetric matrices (Simo & Vu-Quoc, 1986)

$$[\boldsymbol{S}(\boldsymbol{a})\boldsymbol{S}(\boldsymbol{b})-\boldsymbol{S}(\boldsymbol{b})\boldsymbol{S}(\boldsymbol{a})]\,\boldsymbol{c} = [\boldsymbol{a}\times\boldsymbol{b}]\times\boldsymbol{c} \quad \forall\boldsymbol{a},\boldsymbol{b},\boldsymbol{c}\in\mathbb{R}^3. \tag{2.72}$$

The skew-symmetric tensors from Eqs. (2.67), (2.69), and (2.71) can be further simplified. Finally, using this definition, the relations

$$\delta\boldsymbol{\theta}_{\delta\varphi} = \delta\varphi\boldsymbol{g}_1, \tag{2.73}$$

$$\delta\boldsymbol{\theta}_{\delta t} = \boldsymbol{T}_{\delta t}\delta\boldsymbol{t} \;\text{ with }\; \boldsymbol{T}_{\delta t} = \frac{1}{||\boldsymbol{t}||}\left(\boldsymbol{I}-\frac{\boldsymbol{g}_1\otimes\boldsymbol{g}_0}{1+\boldsymbol{g}_1^T\boldsymbol{g}_0}\right)\boldsymbol{S}(\boldsymbol{g}_1), \;\text{ and} \tag{2.74}$$

$$\begin{aligned}\delta\boldsymbol{\theta}_{\delta t_0} &= \boldsymbol{T}_{\delta t_0}\delta\boldsymbol{t}_0 \;\text{ with}\\ \boldsymbol{T}_{\delta t_0} &= \frac{1}{||\boldsymbol{t}_0||}\left[-\boldsymbol{S}(\boldsymbol{g}_1)+\frac{(\boldsymbol{g}_1\otimes\boldsymbol{g}_1)\boldsymbol{S}(\boldsymbol{g}_0)-(\boldsymbol{g}_0\times\boldsymbol{g}_1)\otimes(\boldsymbol{g}_0+\boldsymbol{g}_1)}{1+\boldsymbol{g}_0^T\boldsymbol{g}_1}\right]\end{aligned} \tag{2.75}$$

are determined. It is pointed out that the relations shown in Eqs. (2.73) and (2.74) are identical to the ones derived in Meier et al. (2014, 2017) by a different approach. Moreover, analogous to Eq. (2.60), $\delta\boldsymbol{\theta}$ can be split into

$$\delta\boldsymbol{\theta} = \delta\boldsymbol{\theta}_{\parallel}+\delta\boldsymbol{\theta}_{\perp} = \delta\Theta_1\boldsymbol{g}_1+\delta\boldsymbol{\theta}_{\perp}, \tag{2.76}$$

where

$$\boldsymbol{g}_1^T\delta\boldsymbol{\theta} = \delta\Theta_1 =: \delta\Theta_{sr}+\delta\varphi \tag{2.77}$$

is the total torsional variation, see Eq. (2.37), and

$$\boldsymbol{t} \times \delta \boldsymbol{t} = \frac{\boldsymbol{t}}{||\boldsymbol{t}||} \times \delta \left(\frac{\boldsymbol{t}}{||\boldsymbol{t}||} \right) = \frac{\boldsymbol{t} \times \delta \boldsymbol{t}}{||\boldsymbol{t}||^2} \tag{2.78}$$

is the part of the rotation vector variation that is perpendicular to $\boldsymbol{g}_1$. Furthermore,

$$\begin{aligned}
\delta \Theta_{sr} &= (\boldsymbol{\Lambda}_{sr}(\boldsymbol{g}_0, \boldsymbol{g}_1)\boldsymbol{\Lambda}_r \boldsymbol{e}_2)^T \delta (\boldsymbol{\Lambda}_{sr}(\boldsymbol{g}_0, \boldsymbol{g}_1)\boldsymbol{\Lambda}_r \boldsymbol{e}_3) \\
&= (\boldsymbol{\Lambda}_{sr}(\boldsymbol{g}_0, \boldsymbol{g}_1)\boldsymbol{\Lambda}_r \boldsymbol{e}_3)^T \delta (\boldsymbol{\Lambda}_{sr}(\boldsymbol{g}_0, \boldsymbol{g}_1)\boldsymbol{\Lambda}_r \boldsymbol{e}_2) \\
&= \boldsymbol{g}_1^T \left(\boldsymbol{T}_{\delta t} \delta \boldsymbol{t} + \boldsymbol{T}_{\delta t_0} \delta \boldsymbol{t}_0 \right)
\end{aligned} \tag{2.79}$$

is the part of the total torsional variation due to the smallest-rotation contribution and $\delta\varphi$ is then the difference to the total torsional variation.

To obtain the linearizations, the operators $\delta(\bullet)$ and $\Delta(\bullet)$ are simply interchanged.

Angular Velocity Vector: The material angular velocity vector in dependence of $\boldsymbol{t}$ and φ reads

$$\boldsymbol{S}(\boldsymbol{W}) = \boldsymbol{\Lambda}^T \dot{\boldsymbol{\Lambda}} = \boldsymbol{\Lambda}^T (\dot{\boldsymbol{R}} \boldsymbol{\Lambda}_{sr} \boldsymbol{\Lambda}_r + \boldsymbol{R} \dot{\boldsymbol{\Lambda}}_{sr} \boldsymbol{\Lambda}_r), \tag{2.80}$$

where it is noted that $\dot{\boldsymbol{t}}_0 = 0$ and

$$\dot{\boldsymbol{g}}_1 = \frac{\dot{\boldsymbol{t}}}{||\boldsymbol{t}||} - \frac{\boldsymbol{t}^T \dot{\boldsymbol{t}}}{||\boldsymbol{t}||^3} \boldsymbol{t}. \tag{2.81}$$

The quantities $\dot{\boldsymbol{\Lambda}}_{sr}$ and $\dot{\boldsymbol{R}}$ are not explicitly given here, since they are in general analogous to Eqs. (2.58) and (2.59). But a short version emerges as

$$\boldsymbol{w} = \boldsymbol{T}_{\delta t} \dot{\boldsymbol{t}} + \dot{\varphi} \boldsymbol{g}_1 \quad \text{and} \quad \boldsymbol{a} = \dot{\boldsymbol{T}}_{\delta t} \dot{\boldsymbol{t}} + \boldsymbol{T}_{\delta t} \ddot{\boldsymbol{t}} + \dot{\varphi} \dot{\boldsymbol{g}}_1 + \ddot{\varphi} \boldsymbol{g}_1. \tag{2.82}$$

Alternatively, $\boldsymbol{w}$ and $\boldsymbol{a}$ can be expressed in terms of $\boldsymbol{t}$ and W_1. To this end, $\boldsymbol{w}$ is split to yield

$$\boldsymbol{w} = \boldsymbol{w}_{\parallel} + \boldsymbol{w}_{\perp} = W_1 \boldsymbol{g}_1 + \frac{\boldsymbol{t} \times \dot{\boldsymbol{t}}}{||\boldsymbol{t}||^2} \tag{2.83}$$

with the according spatial angular acceleration vector reading

$$\boldsymbol{a} = \dot{W}_1 \boldsymbol{g}_1 + W_1 \dot{\boldsymbol{g}}_1 + \boldsymbol{g}_1 \times \ddot{\boldsymbol{g}}_1. \tag{2.84}$$

φ and W_1 are related via

$$W_1 = W_{\mathrm{sr}} + \dot{\varphi} \tag{2.85}$$

with

$$\begin{aligned}
W_{\mathrm{sr}} &= (\boldsymbol{\Lambda}_{sr}(\boldsymbol{g}_0, \boldsymbol{g}_1)\boldsymbol{\Lambda}_r \boldsymbol{e}_2)^T (\dot{\boldsymbol{\Lambda}}_{sr}(\boldsymbol{g}_0, \boldsymbol{g}_1)\boldsymbol{\Lambda}_r \boldsymbol{e}_3) \\
&= (\boldsymbol{\Lambda}_{sr}(\boldsymbol{g}_0, \boldsymbol{g}_1)\boldsymbol{\Lambda}_r \boldsymbol{e}_3)^T (\dot{\boldsymbol{\Lambda}}_{sr}(\boldsymbol{g}_0, \boldsymbol{g}_1)\boldsymbol{\Lambda}_r \boldsymbol{e}_2).
\end{aligned} \tag{2.86}$$

Update procedure: Following the interpolation scheme presented in this section, the increments $\Delta \boldsymbol{t}$ and $\Delta\varphi$ will result from the iterative solution scheme. These increments could be converted to $\Delta\theta$ using Eqs. (2.73)–(2.75) to update $\boldsymbol{\Lambda}$ according to Eq. (2.40). However, this approach did not prove to be favorable. Instead the nodal rotation tensors should be updated using

$$\boldsymbol{g}^{k+1} = \frac{\boldsymbol{g}^k + \Delta \boldsymbol{t}^{k+1}}{||\boldsymbol{g}^k + \Delta \boldsymbol{t}^{k+1}||} \tag{2.87}$$

according to

$$\boldsymbol{\Lambda}^{k+1} = \boldsymbol{R}(\boldsymbol{g}^{k+1}, \Delta\varphi^{k+1})\boldsymbol{\Lambda}_{sr}(\boldsymbol{g}^k, \boldsymbol{g}^{k+1})\boldsymbol{\Lambda}^k. \tag{2.88}$$

2.3.3.3 Justification of Choice of Rotation Parameterization and Interpolation

First of all, possible interpolation schemes were chosen from the large body of literature on geometrically exact beam elements disregarding work on fundamentally different modeling approaches such as degenerated formulations (Bathe & Bolourchi, 1979; Dvorkin et al., 1988) and co-rotational formulations (Battini & Pacoste, 2002; Belytschko & Hsieh, 1973; Belytschko & Glaum, 1979; Crisfield, 1990; Wempner, 1969). Many contributions on geometrically exact beams are concerned with the parameterization and interpolation of the rotations (Betsch & Steinmann, 2002; Crisfield & Jelenić, 1999; Crivelli & Felippa, 1993; Ghosh & Roy, 2008, 2009; Ibrahimbegović et al., 1995; Jelenić & Crisfield, 1998, 1999; Jelenić & Saje, 1995; Petrov & Géradin, 1998; Romero & Armero, 2002; Romero, 2004; Sansour & Wagner, 2003). Here, it should be noted that Crisfield & Jelenić (1999) and Jelenić & Crisfield (1999) were the first to propose an objective and path-independent orthogonal interpolation scheme.

Secondly, the choice was further limited based on the work by Romero (2004) and Romero (2008a) who performed comparative studies. Their results indicate that geometrically exact beams in addition with the orthogonal interpolation strategy proposed by Jelenić & Crisfield (1999) are favorable in terms of accuracy, robustness, and efficiency. Unfortunately, the smallest-rotation interpolation scheme from Sect. 2.3.3.2 were not part of the study.

Thirdly, Meier et al. (2017) compared different Kirchhoff-Love beam formulations. These formulations differed in the manner the Kirchhoff constraints were enforced (in a strong or weak manner) and they differed in their respective nodal triad parameterization. Meier et al. (2017) concluded that the parameterization based on nodal rotation vectors leads to worse Newton-Raphson performance than a parameterization based on a tangent and a scalar torsional parameter. As will be seen in Sect. 3.6, the current formulation results in more favorable behavior when using rotations vectors.

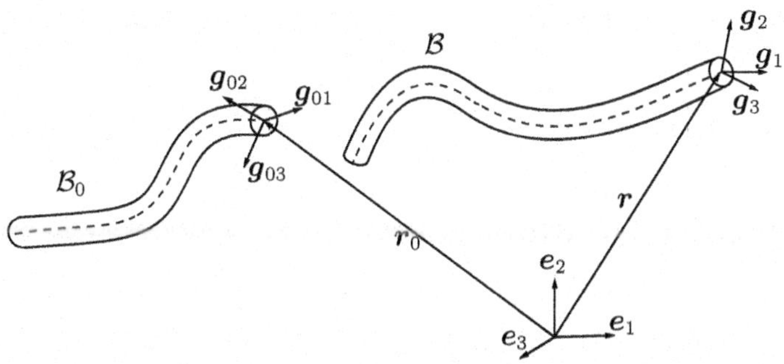

Fig. 2.2 Kinematic quantities defining the reference and current configurations $\mathcal{B}_0$ and $\mathcal{B}$ of Simo-Reissner beams, respectively[1]

2.4 Simo-Reissner Beam Formulation

2.4.1 Kinematic Assumptions and Description

Acquiescing slight redundancies with Sects. 2.2 and 2.3, the kinematics of the Simo-Reissner beam are specified. Generally, it obeys the restriction of rigid cross-sections formulated in Eq. (2.20). Thus, following Eq. (2.23), the current and material configuration $\mathcal{B}$ and $\mathcal{B}_0$ are uniquely defined by $(\boldsymbol{r}(s_0, t), \boldsymbol{\Lambda}(s_0, t))$ and $(\boldsymbol{r}_0(s_0, t), \boldsymbol{\Lambda}_0(s_0, t))$, respectively. Since orthogonal tenors can be defined by at least three parameters, see Sect. 2.3.3, a total of at least 6 degrees of freedom point-wise along the beam are required to describe any configuration. It is noted that all quantities are given in terms of the material curvilinear coordinate or the material arc-length parameter $s_0 \in [0, L_0] \subset \mathbb{R}$ with L_0 being the initial beam length. $\boldsymbol{\Lambda}$ and $\boldsymbol{\Lambda}_0$ define the moving spatial and material frames $\boldsymbol{g}_1, \boldsymbol{g}_2, \boldsymbol{g}_3$ and $\boldsymbol{g}_{01}, \boldsymbol{g}_{02}, \boldsymbol{g}_{03}$, respectively, with respect to a fixed frame $\boldsymbol{e}_1, \boldsymbol{e}_2, \boldsymbol{e}_3$. The kinematic quantities are schematically presented in Fig. 2.2.

Arbitrarily curved reference configurations can be accommodated in the present kinematical description, but beams with an (arbitrary) cross-section, which is constant throughout their length are required. An extension to a case where the center of shear is not identical to the cross-section centroid is not addressed. However, it can be realized by standard methods available in the literature, see e.g. Jafari & Mahjoob (2010). Finally, it is pointed out that s_0 is not an arc-length parameterization of the current configuration because $||\boldsymbol{r}'(s_0)|| \neq 1$.

[1] Figure reprinted with permission from Schulz & Böl (2019). ©2019 by Springer Nature

2.4.2 Balance of Linear and Angular Momentum

A material segment (infinitesimal subset) of a deformed rod

$$\mathcal{B}_m := \bigcup_{s_0 \in \mathcal{I}} \mathcal{B}(s_0) \tag{2.89}$$

with $\mathcal{I} := [s_0^-, s_0^+]$ is considered. On the cross-sections $\mathcal{B}(s_0^-)$ and $\mathcal{B}(s_0^+)$ confining the subset, the resultant contact forces $\boldsymbol{n}^-$ and $\boldsymbol{n}^+$ and the resultant contact couples $\boldsymbol{m}^-$ and $\boldsymbol{m}^+$ act, respectively. Additionally, all other forces (external and body forces) per unit reference length and body couple per unit reference length are depicted by $\boldsymbol{f}$ and $\boldsymbol{l}$, respectively. Moreover, ρ_0 is the mass density per reference unit and per reference cross-section area A. The relevant kinematic quantities and loads are depicted in Fig. 2.3.

Balance of linear momentum: The resultant force on the material segment is equal to the time derivative of the linear momentum such that

$$\boldsymbol{n}^+(s_0^+, t) - \boldsymbol{n}^-(s_0^-, t) + \int_{s_0^-}^{s_0^+} \boldsymbol{f}(s_0, t)\mathrm{d}s_0 = \frac{\mathrm{d}}{\mathrm{d}t} \int_{s_0^-}^{s_0^+} \left(\int_{\mathcal{B}(s_0)} \rho(s_0)\dot{\boldsymbol{x}}(x, t)\mathrm{d}x_2\mathrm{d}x_3 \right) \mathrm{d}s_0. \tag{2.90}$$

The constrained motion defined in Eq. (2.20) in combination with the assumptions that $\boldsymbol{r}$ runs through all centroids simplifies the integral over the rigid cross-section according to

$$\int_{\mathcal{B}(s_0)} \rho(s_0)\dot{\boldsymbol{x}}(x, t)\mathrm{d}x_2\mathrm{d}x_3 = (\rho_0 A)(s_0)\dot{\boldsymbol{r}}(s_0, t), \tag{2.91}$$

since the first moment of area vanishes. Then, $\rho_0 A$ is the reference mass density per reference length. Letting $s^- \to s^+$, assuming continuity of all involved quantities such that

$$\boldsymbol{n}^+(a, t) = \boldsymbol{n}^-(a, t) \quad \forall a \in [0, L_0], \tag{2.92}$$

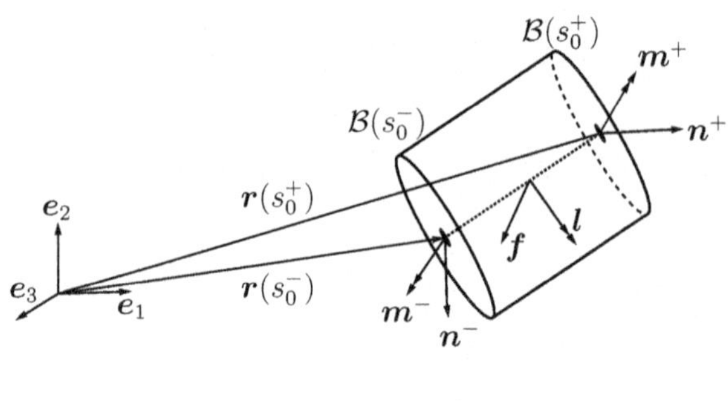

Fig. 2.3 Schematic depiction of loads and moments acting on an infinitesimal element of a beam

and dividing Eq. (2.90) by $\mathrm{d}s_0$ yields the first half of the classic form

$$\boldsymbol{n}'(s_0, t) + \boldsymbol{f}(s_0, t) = (\rho_0 A)(s_0)\ddot{\boldsymbol{r}}(s_0, t) \quad \text{for} \quad s_0 \in [0, L_0], t \geq 0. \tag{2.93}$$

Balance of angular momentum: Here, the sum of the resultant couples on the material segment is equal to the time derivative of the angular momentum. At this, the angular momentum is determined with respect to the origin of the coordinate system such that

$$\begin{aligned}
&\boldsymbol{m}^+(s_0^+, t) - \boldsymbol{m}^-(s_0^-, t) + \boldsymbol{r}(s_0^+, t) \times \boldsymbol{n}^+(s_0^+, t) - \boldsymbol{r}(s_0^-, t) \times \boldsymbol{n}^-(s_0^-, t) \\
&+ \int_{s_0^-}^{s_0^+} (\boldsymbol{r}(s_0, t) \times \boldsymbol{f}(s_0, t) + \boldsymbol{l}(s_0, t))\, \mathrm{d}s_0 \\
&= \frac{\mathrm{d}}{\mathrm{d}t} \int_{s_0^-}^{s_0^+} \left(\int_{\mathcal{B}(s_0)} \rho(s_0)(\boldsymbol{x}(x, t) \times \dot{\boldsymbol{x}}(x, t))\mathrm{d}x_2\mathrm{d}x_3 \right) \mathrm{d}s_0.
\end{aligned} \tag{2.94}$$

Again, the integral over the rigid cross-section can be simplified due to the fact that $\boldsymbol{x}$ is constrained according to Eq. (2.20), that $\boldsymbol{r}$ runs through all centroids, that $\boldsymbol{g}_2$ and $\boldsymbol{g}_3$ are the principal axis of inertia, and that the first momentum of area vanishes. Consequently,

$$\int_{\mathcal{B}(s_0)} \rho(s_0)(\boldsymbol{x}(x, t) \times \dot{\boldsymbol{x}}(x, t))\mathrm{d}x_2\mathrm{d}x_3 = (\rho_0 A)(s_0)(\boldsymbol{r}(s_0, t) \times \dot{\boldsymbol{r}}(s_0, t)) + \boldsymbol{c}_\rho(s_0, t)\boldsymbol{w}(s_0, t), \tag{2.95}$$

where based on Eq. $(2.33)_4$ the relation

$$\dot{\boldsymbol{g}}_i(s_0, t) = \boldsymbol{w}(s_0, t) \times \boldsymbol{g}_i(s_0, t) \tag{2.96}$$

is used. With I_2 and I_3 being the two second moments of area about $\boldsymbol{g}_2$ and $\boldsymbol{g}_3$, respectively, the spatial inertia tensor per unit reference

$$\boldsymbol{c}_\rho(s_0, t) = \rho_0(s_0) \cdot \mathrm{diag}[(I_2(s_0) + I_3(s_0)), I_2(s_0), I_3(s_0)]_{\boldsymbol{g}_i(s_0, t)} \tag{2.97}$$

is identified. Assuming continuity of all quantities, incorporating Eqs. (2.95) and (2.93) into Eq. (2.94), and dividing Eq. (2.94) by $\mathrm{d}s_0$ yields the second half of the classic form

$$\begin{aligned}
\boldsymbol{m}'(s_0, t) + \boldsymbol{r}'(s_0, t) \times \boldsymbol{f} + \boldsymbol{l} &= \frac{\mathrm{d}}{\mathrm{d}t} \left(\boldsymbol{c}_\rho(s_0, t)\boldsymbol{w}(s_0, t)\right) \\
&= \boldsymbol{S}(\boldsymbol{w}(s_0, t))\boldsymbol{c}_\rho(s_0, t)\boldsymbol{w}(s_0, t) + \dot{\boldsymbol{w}}(s_0, t) \quad \text{for} \quad s_0 \in [0, L_0], t \geq 0.
\end{aligned} \tag{2.98}$$

A material version of the strong from is attainable involving the material contact forces and contact couples $\boldsymbol{N} := \boldsymbol{\Lambda}^T \boldsymbol{n}$ and $\boldsymbol{M} := \boldsymbol{\Lambda}^T \boldsymbol{m}$, respectively. Also, the spatial and material angular acceleration vectors $\boldsymbol{a} := \dot{\boldsymbol{w}}$ and $\boldsymbol{A} := \dot{\boldsymbol{W}}$ are defined, respectively.

2.4.3 Weak Form and Resulting Deformation Measures

For convenience, the strong form resulting from the balance of linear and angular momentum presented in Sect. 2.4.2 is repeated here shortened and with the arguments dropped as

$$\boldsymbol{n}' + \boldsymbol{f} + \boldsymbol{f}_\rho = \boldsymbol{0} \tag{2.99}$$

with $\boldsymbol{f}_\rho := -\rho A \ddot{\boldsymbol{r}}$ and

$$\boldsymbol{m}' + \boldsymbol{l} + \boldsymbol{r}' \times \boldsymbol{n} + \boldsymbol{m}_\rho = \boldsymbol{0}, \tag{2.100}$$

where $\boldsymbol{m}_\rho := -\boldsymbol{S}(\boldsymbol{w})\boldsymbol{c}_\rho \boldsymbol{w} - \boldsymbol{c}_\rho \boldsymbol{a} = -\boldsymbol{\Lambda}[\boldsymbol{S}(\boldsymbol{W})\boldsymbol{C}_\rho \boldsymbol{W} + \boldsymbol{C}_\rho \boldsymbol{A}]$. Herein,

$$\boldsymbol{C}_\rho = \boldsymbol{\Lambda}^T \boldsymbol{c}_\rho \boldsymbol{\Lambda} \tag{2.101}$$

is the material inertia tensor per unit reference. In order to complete an initial boundary problem, which is generally referred to as *strong form*, the boundary conditions on the Dirichlet and Neumann boundary Γ_u and Γ_σ, respectively, are defined to be

$$\boldsymbol{r} = \bar{\boldsymbol{r}}, \boldsymbol{\Lambda} = \bar{\boldsymbol{\Lambda}} \text{ on } \Gamma_u \text{ and } \boldsymbol{n} = \bar{\boldsymbol{n}},\ \boldsymbol{m} = \bar{\boldsymbol{m}} \text{ on } \Gamma_\sigma \tag{2.102}$$

and the initial conditions

$$\boldsymbol{r} = \boldsymbol{r}_0,\ \dot{\boldsymbol{r}} = \boldsymbol{v}_0,\ \boldsymbol{\Lambda} = \boldsymbol{\Lambda}_0,\ \boldsymbol{w} = \boldsymbol{w}_0 \text{ at } t = 0. \tag{2.103}$$

Furthermore, $\Gamma_u \cap \Gamma_\sigma = \varnothing$ and $\Gamma_u \cup \Gamma_\sigma = \{0, L_0\}$ hold.

In the remainder of this section, the weak form is devised. To this end, the *method of weighted residuals* is implemented. For mathematical details of this method and its mechanical interpretations as either *principle of virtual work* or *principle of minimum of total potential energy* the reader is referred to Bathe (2006), Hughes (2012), Zienkiewicz & Taylor (2005) and Zienkiewicz et al. (2005). According to these methods, the admissible variations need to be defined first. These infinitesimal small and arbitrary changes of $\mathcal{B}$ are depicted by $\delta\mathcal{B}$ and described by

$$[0, L_0] \ni s_0 \mapsto \delta\boldsymbol{r}(s_0),\ \delta\boldsymbol{\theta}(s_0) \in \mathbb{R}^3 \times \mathbb{R}^3, \tag{2.104}$$

where $\delta\boldsymbol{r}$ is the virtual displacement vector and $\delta\boldsymbol{\theta}$ depicts the virtual spatial rotation vector as already introduced in Eq. $(2.33)_2$. Secondly, the strong form given in Eqs. (2.99) and (2.100) are multiplied by $\delta\boldsymbol{r}$ and $\delta\boldsymbol{\theta}$, respectively, and integrated over the whole domain. Subsequent partial integration results in

$$G_{SR} = \int_0^{L_0} \left((\delta r' - \delta\theta \times r')^T n + \delta\theta'^T m - \delta r^T (f + f_\rho) - \delta\theta^T (l + m_\rho) \right) \mathrm{d}s_0$$
$$- [\delta r^T \bar{n} + \delta\theta^T \bar{m}]_{\Gamma_\sigma} \doteq 0, \tag{2.105}$$

which is designated the *weak form* in combination with the restriction

$$\delta r = \delta\theta = 0 \ \text{ on } \ \Gamma_u \tag{2.106}$$

and the conditions formulated in Eqs. (2.102) and (2.103). To conclude the mathematical foundation, as usual, a trial space $\mathcal{T}_{\mathrm{SR}} \ni r(s_0, t), \Lambda(s_0, t)$ of all functions with square-integrable derivatives of first order, which also complies with the above mentioned boundary conditions, is introduced in combination with a weighting space $\mathcal{W}_{\mathrm{SR}} \ni \delta r(s_0), \delta\theta(s_0)$ of all functions with square-integrable derivatives of first order also complying with Eq. (2.106). According to the principal of virtual work in Eq. (2.105) n and m are paired with the objective or corrotated variations

$$\delta_{\mathrm{rel}}\gamma = \delta r' - \delta\theta \times r' \ \text{ and } \ \delta_{\mathrm{rel}}\omega = \delta\theta' \tag{2.107}$$

of the spatial deformation measures γ and ω, respectively. For an arbitrary vector $a \in \mathbb{R}^3$, $\delta_{\mathrm{rel}} a := \delta a - \delta\theta \times r'$ constitutes the objective part of the variation and defines the relative variation of a with respect to an observer that moves according to $\delta\theta$. The original quantities belonging to the variations given in Eq. (2.107) can be shown to be (Simo & Wong, 1991)

$$\gamma = r' - g_1 \ \text{ and } \ \omega = k - \Lambda\Lambda_0^T k_0. \tag{2.108}$$

Conversely, the same procedure performed so far can be started at the material formulation of the strong form, see e.g. Crisfield & Jelenić (1999), yielding the material deformation vectors

$$\Gamma = \Lambda^T r' - E_1 \ \text{ and } \ \Omega = K - K_0 \tag{2.109}$$

and the respective variations

$$\delta\Gamma = \Lambda^T (\delta r' - \delta\theta \times r') \ \text{ and } \ \delta\Omega = \Lambda^T \delta\theta'. \tag{2.110}$$

As required for meaningful deformation measures, they all vanish in an undeformed state, i.e. when $\Lambda = \Lambda_0$ and $r' = g_1 = r_0'$. A second important quality of meaningful deformation measures is objectivity, i.e. the invariance of these measures under a rigid body transformation. A proof is given in the appendix of Crisfield & Jelenić (1999).

2.4.4 Constitutive Relations

Simple linear constitutive dependencies of the material stress resultants $\boldsymbol{N}$ and $\boldsymbol{M}$ on the deformation measures $\boldsymbol{\Gamma}$ and $\boldsymbol{\Omega}$ are introduced (Simo & Vu-Quoc, 1986) as

$$\boldsymbol{N} = \frac{\partial \Pi_{\text{int}}}{\partial \boldsymbol{\Gamma}} = \boldsymbol{C}_N \text{ and } \boldsymbol{M} = \frac{\partial \Pi_{\text{int}}}{\partial \boldsymbol{\Omega}} = \boldsymbol{C}_M \boldsymbol{\Omega} \tag{2.111}$$

with

$$\Pi_{\text{int}}(\boldsymbol{\Gamma}, \boldsymbol{\Omega}) = \frac{1}{2}\boldsymbol{\Gamma}^T \boldsymbol{C}_N \boldsymbol{\Gamma} + \frac{1}{2}\boldsymbol{\Omega}^T \boldsymbol{C}_M \boldsymbol{\Omega}, \tag{2.112}$$

being the length-specific internal energy. Herein, $\boldsymbol{C}_N = \textbf{diag}[C_{N1}, C_{N2}, C_{N3}]_{\boldsymbol{E}_i}$ and $\boldsymbol{C}_M = \textbf{diag}[C_{M1}, C_{M2}, C_{M3}]_{\boldsymbol{E}_i}$ are diagonal matrices featuring the constitutive constants, which can be measured experimentally or derived theoretically. In the simplest case,

$$C_{N1} = EA, \quad C_{N2} = G\bar{A}_2, \text{ and } C_{N3} = G\bar{A}_3 \tag{2.113}$$

and

$$C_{M1} = GI_p, \quad C_{M2} = EI_2, \text{ and } C_{M3} = EI_3 \tag{2.114}$$

with E to be Young's modulus, A is the cross-section area, G is the shear modulus, I_p depicts the polar moment of inertia, $\bar{A}_2$ and $\bar{A}_3$ are the reduced cross-sections, and I_2 as well as I_3 are the two principle moments of inertia.

Arbitrary functions of Π_{int} in terms of $\boldsymbol{\Gamma}$ and $\boldsymbol{\Omega}$ can easily be implemented, but an extension to hyper-elastic cross-sectional material behavior is more involved and not addressed here, see e.g. Wackerfuß & Gruttmann (2009) and the references therein.

Furthermore, the one-dimensional constitutive laws presented in this section can be derived from a three-dimensional approach under the assumption of small local strains, see e.g. Meier et al. (2017).

2.5 Kirchhoff-Love Beam Formulations

The Simo-Reissner beam formulation presented in Sect. 2.4 has 6 degrees of freedom everywhere along its center-line. In some applications, it might be sensible to employ a shear-free, so-called Kirchhoff-Love beam formulation (Kirchhoff, 1859; Love, 2013) offering certain advantages specified in Sect. 2.5.1. The kinematic constraint of vanishing shear deformations, the so-called Kirchhoff constraint, can be cast as

$$\boldsymbol{g}_1(s_0) \equiv \boldsymbol{r}'(s_0)/||\boldsymbol{r}'(s_0)|| \text{ or } \boldsymbol{r}'(s_0) \cdot \boldsymbol{g}_2(s_0) \equiv 0 \text{ and } \boldsymbol{r}'(s_0) \cdot \boldsymbol{g}_3(s_0) \equiv 0, \tag{2.115}$$

which can be enforced in a weak manner involving Lagrangian multipliers, see Sect. 2.5.4, as well as in a strong manner effectively reducing the degrees of freedom.

Furthermore, the slenderness ratio $\zeta := L_0/R_{cs}$ is defined to specify the extent of slenderness of a body, where R_{cs} is a characteristic value of the cross-section, e.g. the side length of a quadratic profile or the diameter of a circular cross-section, which is assumed in the remainder.

2.5.1 Advantages of Kirchhoff-Love Beam Formulations

The Kirchhoff-Love beam formulation may be employed when simulations involve very slender components such as e.g. single DNA chains (Yang et al., 1993), woven fabrics (Durville, 2010), Nickel Titanium wires (Gadot et al., 2015), and cellulose nanopaper (Kulachenko et al., 2012). Naturally, excluding certain deformation modes leads to a deviation from fully three-dimensional results (e.g. Borri et al., 1992; Giavotto et al., 1983; Han & Bauchau, 2016) but Meier et al. (2014) found that the neglect of shear deformation in structures with $\zeta = 100$ exhibit a model error on the order of 0.01%, only, relative to beam elements including shear deformation. In comparison to other beam formulations that include shear deformations, the Kirchhoff-Love beam theory offers several numerical advantages, which were comprehensively studied by Meier et al. (2017) and are mentioned here as well.

First of all, obviously, when enforcing the Kirchhoff constraint in a strong manner, the total number of required degrees of freedom along the center-line will be reduced. Still, it remains to examine if the same convergence rates and discretization error levels are sustained. But since Meier et al. (2017) found these properties not deteriorated in their approach to modeling Kirchhoff-Love beams, it is also not expected here. This anticipation is confirmed in Sect. 3.6.

Secondly, it is well-known that high differences in stiffnesses, which are present in highly slender bodies, result in ill-conditioned tangent matrices exhibiting an extreme ration of highest (from shear or axial deformation modes) to lowest (from torsional or bending deformation modes) eigenvalues. The involved matrices are often characterized by a condition number. For linear and nonlinear solvers, bad condition numbers might lead to a deterioration of the numerical solver, which is especially unfavorable in dynamic problems where errors can accumulate. To illustrate the problem, an initially straight beam with $L_0 = 1$ and a circular cross-section can be considered. Then, the contributions to the tangent matrix of bending, torsion, axial, and shear deformation modes κ_b, κ_t, κ_a, and κ_s, respectively, show the proportionalities

$$\kappa_b \sim EI_2 \sim \zeta^{-4}, \quad \kappa_t \sim GI_p \sim \zeta^{-4}, \quad \kappa_a \sim EA \sim \zeta^{-2}, \quad \kappa_s \sim G\bar{A} \sim \zeta^{-2}. \tag{2.116}$$

It becomes evident that when the length is held constant but the slenderness ratio ζ increases, κ_b and κ_t decrease distinctively faster than κ_a and κ_s. But this finding also implies that excluding shear deformation does not completely remedy the deficiencies. Axial deformation would also need to be removed, which will be tested in Sect. 3.6. It is found that excluding shear deformation modes vastly improves the

Newton-Raphson performance, but removing axial deformation modes did not alter the performance.

Thirdly, high eigenfrequencies arise for very slender beams, i.e. the associated partial differential equation is stiff. Thus, in explicit time integration schemes, very small time steps become necessary to meet the stability requirements although often only a temporal resolution covering the low eigenfrequencies would be required. Conversely, implicit solution variants produce unconditional stability in the small deformation domain. But for finite deformations, again, high frequencies lead to high discretization error levels and poor numerical performance. Thus, it would be favorable to exclude such high frequencies. Proportionality relations between pure bending, torsion, axial, and shear deformation eigenfrequencies ω_b, ω_t, ω_a, and ω_s, respectively, and geometric quantities are found to be

$$\omega_b \sim \sqrt{\frac{E}{\zeta^2 L_0^2 \rho_0}}, \quad \omega_t \sim \sqrt{\frac{E}{L_0^2 \rho_0}}, \quad \omega_a \sim \sqrt{\frac{E}{L_0^2 \rho_0}}, \quad \text{and} \quad \omega_s \sim \sqrt{\frac{\zeta^2 E}{L_0^2 \rho_0}}. \tag{2.117}$$

Theoretically, it can be easily deduced from these relations that removing shear deformation modes could improve numerical performance. But restricting axial modes seems only to be promising when excluding torsion as well. However, these effects are not studied numerically here, but appear to be interesting for further studies.

2.5.2 *Kirchhoff-Love Beam Formulation Based on Constraint Translation*

2.5.2.1 Kinematic Assumptions and Description

As with the Simo-Reissner beam in Sect. 2.4, an arbitrarily curved reference configuration of the beam whose line through all cross-section centroids is described by the position vector $\boldsymbol{r}_0(s_0)$ is considered, where $s_0 \in [0, L_0] \subset \mathbb{R}$ is the arc-length coordinate and L_0 is the initial length of the axis of the beam. The basic kinematic quantities are depicted in Fig. 2.4.

Again, besides the inertial Cartesian frame composed of $\boldsymbol{e}_1$, $\boldsymbol{e}_2$, and $\boldsymbol{e}_3$, a right-handed orthonormal triad field attached to the beam with the base vectors $\boldsymbol{g}_{01}(s_0)$, $\boldsymbol{g}_{02}(s_0)$, and $\boldsymbol{g}_{03}(s_0)$ are introduced with the restriction that $\boldsymbol{g}_{01}(s_0)$ is tangent to $\boldsymbol{r}_0$ and always coincides with the normal of the arbitrary cross-section at s_0. Of course, $\boldsymbol{g}_{02}(s_0)$ and $\boldsymbol{g}_{03}(s_0)$ are identical to the principle axes of inertia of the cross-section at s_0. Analogously, in the current configuration the current center line $\boldsymbol{r}(s_0)$ and the base vectors $\boldsymbol{g}_1(s_0, t)$, $\boldsymbol{g}_2(s_0, t)$, and $\boldsymbol{g}_3(s_0, t)$ are introduced.

In shear-deformable beam formulations both, the center line position vector field and the orthogonal transformation tensor field are needed to uniquely define a configuration requiring a total of 6 degrees of freedom along the beam. But taking advantage of the Kirchhoff constraint defined in Eq. (2.115) this number can be reduced. To

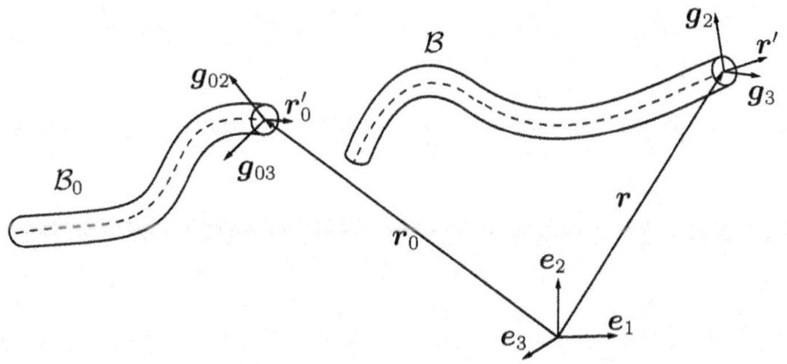

Fig. 2.4 Kinematic quantities defining the reference and current configurations $\mathcal{B}_0$ and $\mathcal{B}$ of Kirchhoff-Love beams, respectively[2]

this end, an axial displacement field

$$u(s_0, t) := s(s_0, t) - s_{0,t} \tag{2.118}$$

is introduced, where s is the arc-length parameter of the deformed beam. Then, if a single arbitrary position vector denoted by $\boldsymbol{r}^* = \boldsymbol{r}(s_0^*)$ is given, the current configuration $\mathcal{B}$ is uniquely defined when also the fields $\boldsymbol{\Lambda}(s_0)$ and $u(s_0)$ are known. Consequently, the current and material configuration $\mathcal{B}$ and $\mathcal{B}_0$ are uniquely defined by $(\boldsymbol{r}^*(t), u(s_0, t), \boldsymbol{\Lambda}(s_0, t))$, and $(\boldsymbol{r}^*(t), \boldsymbol{\Lambda}_0(s_0, t))$, respectively. Following this parameterization of the Kirchhoff-Love beam, the derivative of the current center line

$$\boldsymbol{r}'(s_0) = (1 + u'(s_0))\boldsymbol{\Lambda}(s_0)\boldsymbol{e}_1 \tag{2.119}$$

is uniquely defined, too, and thus $\boldsymbol{r}$ is given implicitly. Summarizing, the new parameterization presented in this Section enforces the Kirchhoff constraint in a strong manner and has 3 rotational degrees and one translational degree of freedom everywhere along the beam, with the exception at $s_0 = s_0^*$. There, a total of 6 parameters are required to define $\boldsymbol{\Lambda}$ and $\boldsymbol{r}^*$. In other words, the Kirchhoff constraint is used to remove two translational degrees of freedom, which is in opposite to the work by Meier et al. (2014) who removed two rotational degrees of freedom. A detailed numerical study of these two approaches is presented in Sect. 3.6.

Resulting from the implicit definition of the center-line, the current center-line

$$\boldsymbol{r}(s_0, t) = \boldsymbol{r}^*(s_0^*) + \int_{s_0^*}^{s_0} (1 + u'(s_0, t))\boldsymbol{\Lambda}(s_0, t)\boldsymbol{e}_1 \mathrm{d}s_0 \tag{2.120}$$

and its variation

[2] Figure reprinted with permission from Schulz & Böl (2019). ©2019 by Springer Nature

$$\delta \boldsymbol{r} = \delta \boldsymbol{r}^* + \int_{s_0^*}^{s_0} \left(-(1+u')\boldsymbol{S}(\boldsymbol{g}_1)\delta\boldsymbol{\theta} + \delta u' \boldsymbol{g}_1\right) \mathrm{d}s_0 \tag{2.121}$$

are defined implicitly. The respective temporal derivatives of first and second grade are

$$\dot{\boldsymbol{r}} = \dot{\boldsymbol{r}}^* + \int_{s_0^*}^{s_0} [\dot{u}'\boldsymbol{g}_1 + (1+u')\dot{\boldsymbol{g}}_1]\mathrm{d}s_0 = \dot{\boldsymbol{r}}^* + \int_{s_0^*}^{s_0} [\dot{u}'\boldsymbol{g}_1 + (1+u')\boldsymbol{S}(\boldsymbol{w})\boldsymbol{g}_1]\mathrm{d}s_0 \tag{2.122}$$

and

$$\ddot{\boldsymbol{r}} = \ddot{\boldsymbol{r}}^* + \int_{s_0^*}^{s_0} [\ddot{u}'\boldsymbol{g}_1 + 2\dot{u}'\dot{\boldsymbol{g}}_1 + (1+u')\ddot{\boldsymbol{g}}_1]\mathrm{d}s_0$$

$$= \ddot{\boldsymbol{r}}^* + \int_{s_0^*}^{s_0} [\ddot{u}'\boldsymbol{g}_1 + 2\dot{u}'\boldsymbol{S}(\boldsymbol{w})\boldsymbol{g}_1 + (1+u')(\boldsymbol{S}(\boldsymbol{a})\boldsymbol{g}_1 + \boldsymbol{S}(\boldsymbol{w})^2\boldsymbol{g}_1)]\mathrm{d}s_0 \tag{2.123}$$

$$= \ddot{\boldsymbol{r}}^* + \int_{s_0^*}^{s_0} [\boldsymbol{\Lambda}\left[\ddot{u}' + 2\dot{u}'\boldsymbol{S}(\boldsymbol{W}) + (1+u')(\boldsymbol{S}(\boldsymbol{A}) + \boldsymbol{S}(\boldsymbol{W})^2)\right]\boldsymbol{E}_1)]\mathrm{d}s_0, \tag{2.124}$$

respectively.

2.5.2.2 Strong Form, Weak Form, and Resulting Objective Deformation Measures

The balance of linear and angular momentum does not need to be repeated. Instead, the results from Sect. 2.4.2 can be easily adapted to the shear-free case. To this end, $\boldsymbol{n}$ is split into a part parallel ($\boldsymbol{n}_\parallel$) and a part perpendicular ($\boldsymbol{n}_\perp$) to $\boldsymbol{g}_1$ resulting in

$$\boldsymbol{n}_\parallel = (\boldsymbol{g}_1^T\boldsymbol{n})\boldsymbol{g}_1 = n_\parallel \boldsymbol{g}_1 \quad \text{and} \quad \boldsymbol{n}_\perp = \boldsymbol{n} - \boldsymbol{n}_\parallel. \tag{2.125}$$

In order to eliminate $\boldsymbol{n}_\perp$ for the shear-free case, Eq. (2.99) is recast as

$$\boldsymbol{g}_1^T(\boldsymbol{n}' + \boldsymbol{f} + \boldsymbol{f}_\rho) = 0, \tag{2.126}$$

yielding a single differential equation. Equally, Eq. (2.100) is used in combination with Eq. (2.119) to obtain

$$\boldsymbol{m}' + \boldsymbol{l} + \boldsymbol{m}_\rho + (1+u')\boldsymbol{g}_1 \times \boldsymbol{n} = 0, \tag{2.127}$$

where the contribution by $\boldsymbol{n}_\perp$ is still present. But when formulating the weak form integration by parts is performed eliminating this contribution.

The weak form is obtained by standard procedure. First, Eqs. (2.126) and (2.127) are multiplied with δu and $\delta\boldsymbol{\theta}$, respectively and the equations are integrated over the whole domain. Secondly, partial integration is realized by employing $\boldsymbol{n} = -\boldsymbol{f} - \boldsymbol{f}_\rho$ from Eq. (2.99). Thirdly, the expression for the first derivative of the constrained center line variation is incorporated yielding the weak form of the Kirchhoff-Love beam based on constrained translation

$$G_{\text{KL-CT}} = \int_0^{L_0} \left(\delta u' n_{||} + \delta\boldsymbol{\theta}'^T \boldsymbol{m} - \delta\boldsymbol{r}^T (\boldsymbol{f} + \boldsymbol{f}_\rho) - \delta\boldsymbol{\theta}^T (\boldsymbol{l} + \boldsymbol{m}_\rho)\right) \mathrm{d}s_0 - [\delta\boldsymbol{\theta}^T \overline{\boldsymbol{m}} + \delta\boldsymbol{r}^T \overline{\boldsymbol{n}}]_{\Gamma_\sigma}. \tag{2.128}$$

When substituting Eq. (2.121) into (2.128), it becomes obvious that the weak form exhibits a double integral. In the finite-element framework, this fact leads to the requirement of additional assembly steps, as addressed and discussed in Sect. 3.4.4.

Eq. (2.128) could have also been obtained by a different strategy. The constrained variation of the center line from Eq. (2.121) could have been directly incorporated into the weak form of the Simo-Reissner beam, see e.g. Simo (1985). Thus, the adaption of the strong form performed in Sect. 2.5.2.2 could be avoided. In other words, by describing the kinematically admissible perturbations $\delta\mathcal{B}$ of the Kirchhoff-Love beam by

$$[0, L_0] \ni s_0 \mapsto \delta u(s_0),\ \delta\boldsymbol{\theta}(s_0) \in \mathbb{R} \times \mathbb{R}^3, \tag{2.129}$$

the work contribution of the shear stresses vanishes. The same findings were presented by Meier et al. (2014), who, in opposite to the approach pursued in this thesis, constrained $\delta\boldsymbol{\theta}$. Their approach is presented in Sect. 2.5.3.

By applying the work-pairing principle Eq. (2.128), the strain measures for the Kirchhoff-Love beam

$$\varepsilon = u', \quad \boldsymbol{\omega} = \boldsymbol{k} - \boldsymbol{\Lambda}\boldsymbol{\Lambda}_0^T \boldsymbol{k}_0, \quad \text{and} \quad \boldsymbol{\Omega} = \boldsymbol{K} - \boldsymbol{K}_0 \tag{2.130}$$

are identified. ε is an objective scalar strain measure in both configurations. The angular strain measures $\boldsymbol{\omega}$ and $\boldsymbol{\Omega}$ are identical to the ones found for the Simo-Reissner beam in Sect. 2.4.3. Their objectivity is discussed there. For an unstressed state, i.e., for $\boldsymbol{\Lambda} = \boldsymbol{\Lambda}_0$ and $u' = 0$, the identified deformation measures vanish. Furthermore, studying Eqs. (2.121) and (2.128) it can be noticed that only the first derivatives of u and δu appear and they can be readily replaced by ε and $\delta\varepsilon$, respectively. A mixed variational expression of Hellinger-Reissner type follows. Additionally, proper boundary conditions

$$\boldsymbol{r}^* = \overline{\boldsymbol{r}},\ \boldsymbol{\Lambda} = \overline{\boldsymbol{\Lambda}},\ \varepsilon = \overline{\varepsilon} \ \text{and} \ \boldsymbol{n} = \overline{\boldsymbol{n}},\ \boldsymbol{m} = \overline{\boldsymbol{m}} \tag{2.131}$$

are defined at the Dirichlet boundary Γ_u and at the Neumann boundary Γ_σ, respectively. As usual, the restrictions hold that $\Gamma_u \cap \Gamma_\sigma = \varnothing$ and $\Gamma_u \cup \Gamma_\sigma = \{0, L_0\}$.

In a small abuse of notation, the definition of the beam is completed by introducing the trial space $\mathcal{T}_{\text{KL-CT}} \subset \mathbb{R}^3 \times \mathbb{R} \times SO(3) \ni \boldsymbol{r}^*(t), \varepsilon(s_0, t), \boldsymbol{\Lambda}(s_0, t)$ satisfying the above mentioned boundary conditions. In detail, $\mathcal{T}_{\text{KL-CT}}$ consists of functions with square-integrable derivatives of zero and first order, since the derivatives of r^* and ε do not appear in the weak form. The same property is exhibited by the functions from the weighting space $\mathcal{W}_{\text{KL-CT}} \subset \mathbb{R}^3 \times \mathbb{R}^3 \times \mathbb{R} \ni \delta \boldsymbol{r}^*, \delta\boldsymbol{\theta}(s_0), \delta\varepsilon(s_0)$ with $\delta\boldsymbol{r}^* = \delta\boldsymbol{\theta} = \delta\varepsilon = 0$ on Γ_u. It is stressed that $\boldsymbol{r}^*$ is not a function of s_0 and does not have to be considered in the linearization and discretization of static problems. In the weak form the highest derivative is of first order and therefore, this beam formulation only requires C^0-continuity.

2.5.2.3 Constitutive Relations, Modeling of Dirichlet Boundary Conditions, and Variational Formulation

The material stress resultants $\boldsymbol{N}_{||} = n_{||}\boldsymbol{e}_1$ and $\boldsymbol{M}$ are set into a linear relation with the deformation measures ε and $\boldsymbol{\Omega}$ reading

$$n_{||} = \frac{\partial \Pi_{\text{int}}}{\partial \varepsilon} = EA\varepsilon \quad \text{and} \quad \boldsymbol{M} = \frac{\partial \Pi_{\text{int}}}{\partial \boldsymbol{\Omega}} = \boldsymbol{C}_M \boldsymbol{\Omega} \tag{2.132}$$

with the length-specific internal energy

$$\Pi_{\text{int}}(\varepsilon, \boldsymbol{\Omega}) = \frac{1}{2} EA\varepsilon^2 + \frac{1}{2} \boldsymbol{\Omega}^T \boldsymbol{C}_M \boldsymbol{\Omega}. \tag{2.133}$$

All constitutive quantities are defined in Sect. 2.4.4. A drawback of the implicit definition of the center-line is the aggravated implementation of boundary conditions, which are realized by considering a Lagrangian multiplier potential. For example, a contribution due to boundary conditions applied at $s_0 = L_0$ is

$$\Pi_{\text{bc}} = \boldsymbol{\lambda}^T \left(\boldsymbol{r}(L_0) - \boldsymbol{\gamma}\right). \tag{2.134}$$

$\boldsymbol{\lambda}$ denotes Lagrange multipliers, which represent the unknown reactive forces enforcing the boundary condition. Furthermore, $\bar{\boldsymbol{r}}$ is the prescribed position vector. Trivially, the variation

$$\delta\Pi_{\text{bc}} = \delta\boldsymbol{\lambda}^T \left(\boldsymbol{r}(L_0) - \bar{\boldsymbol{r}}\right) + \delta\boldsymbol{r}(L_0)^T \boldsymbol{\lambda} \tag{2.135}$$

and the linearization

$$\Delta\delta\Pi_{\text{bc}} = \delta\boldsymbol{\lambda}^T \Delta\boldsymbol{r}(L_0) + \boldsymbol{\lambda}^T \Delta\delta\boldsymbol{r}(L_0) + \delta\boldsymbol{r}(L_0)^T \Delta\boldsymbol{\lambda} \tag{2.136}$$

follow. For the sake of completeness, the weak form of the new Kirchhoff-Love beam shall be derived following the Hamilton principle. To this end, the Lagrangian functional

$$\mathcal{L}[\boldsymbol{r}, \boldsymbol{\Lambda}](T) := \int_0^T \Pi_{\text{kin}}[\boldsymbol{r}, \boldsymbol{\Lambda}](t) - \Pi_{\text{int}}[\boldsymbol{r}, \boldsymbol{\Lambda}](t) + W_{\text{ext}}[\boldsymbol{r}, \boldsymbol{\Lambda}](t)\mathrm{d}t \tag{2.137}$$

is defined, combining the contributions of kinetic, internal forces, and external forces

$$\Pi_{\text{kin}}[\dot{\boldsymbol{r}}, \boldsymbol{\Lambda}](t) = \int_0^L \left(\frac{1}{2}\boldsymbol{W}(s_0, t)^T \boldsymbol{C}_\rho \boldsymbol{W}(s_0, t) + \frac{1}{2}\rho A \dot{\boldsymbol{r}}^T(s_0, t)\dot{\boldsymbol{r}}(s_0, t)\right) \mathrm{d}s_0, \tag{2.138}$$

$$\Pi_{\text{int}}[\boldsymbol{r}, \boldsymbol{\Lambda}](t) = \int_0^L \left(\frac{1}{2}(\boldsymbol{K}(s_0, t) - \boldsymbol{K}_0(s_0))^T \boldsymbol{C}_M (\boldsymbol{K}(s_0, t) - \boldsymbol{K}_0(s_0)) + \frac{1}{2}EA\varepsilon^2(s_0, t)\right) \mathrm{d}s_0, \tag{2.139}$$

and

$$W_{\text{ext}}[\boldsymbol{r}, \boldsymbol{\Lambda}](t) = \int_0^L \left((\boldsymbol{r}(s_0, t) - \boldsymbol{r}_0(s_0))^T \boldsymbol{f}(s_0, t)\right) \mathrm{d}s_0, \tag{2.140}$$

respectively. Axial moments cannot be included following this approach since they are not path-independent. Thus, it is required to either start with the strong form or to employ rotation dependent moments, see e.g. Teh & Clarke (1997) and Ritto-Corrêa & Camotim (2003) and the many references therein. But this topic is not within the scope of this thesis.

Subsequently, following Hamilton's principle of least action, it is required that the Gâteaux differential of Eq. (2.137) vanishes to find the true motion. Artificially incorporating the contributions of distributed external moments $\boldsymbol{l}$, of the discrete prescribed loads $\bar{\boldsymbol{m}}$ and $\bar{\boldsymbol{n}}$ at the Neumann boundary Γ_σ, and of the prescribed positions $\bar{\boldsymbol{r}}$ and rotations $\bar{\boldsymbol{\Lambda}}$ at the Dirichlet boundary Γ_u yields the weak form already given in Eq. (2.128).

2.5.3 *Kirchhoff-Love Beam Formulation Based on Constraint Rotation*

In Sect. 2.5.2, a new strategy to incorporating the Kirchhoff constraint was presented. In contrast, in this section the approach devised by Meier et al. (2014) is briefly introduced for later reference. In their work, two rotational degrees of freedom are removed by introducing the triads attached to the cross-sections as function of the center-line $\boldsymbol{r}$, its derivatives, and a torsional parameter.

2.5.3.1 Kinematic Assumptions and Description

The same kinematic quantities presented in Sect. 2.5.2.1 and shown in Fig. 2.4 are considered and not repeated here. Again, the motion is confined according to the Eqs. (2.20) and (2.115). Then, removing two rotational degrees of freedom, every configuration is uniquely defined by

$$[0, L_0], \mathbb{R}^+ \ni s_0, t \mapsto \boldsymbol{r}(s_0, t), \ \varphi(s_0, t) \in \mathbb{R}^3 \times \mathbb{R}, \tag{2.141}$$

where φ depicts a local torsional parameter as introduced in Sect. 2.3.3.2. Thus, the current and material configuration $\mathcal{B}$ and $\mathcal{B}_0$ are uniquely defined by $(\boldsymbol{r}(s_0, t), \varphi(s_0, t))$ and $(\boldsymbol{r}_0(s_0, t), \varphi_0(s_0, t))$, respectively.

A crucial problem of this approach is a definition of the triad attached to the cross-section, which is also objective and applicable to a straight beam, i.e. the second derivative of $\boldsymbol{r}$ must not be involved in the denominator. The classic Frenet-Serret frame for example does not meet these requirements. Therefore, Meier et al. (2014) defined the triads to be based on the smallest-rotation strategy presented in Eq. (2.51) such that

$$\boldsymbol{g}_i = \boldsymbol{R}(\boldsymbol{g}_1, \varphi)\boldsymbol{\Lambda}_{sr}(\boldsymbol{g}_0, \boldsymbol{g}_1)\boldsymbol{\Lambda}_r \boldsymbol{e}_i \ \text{ with } \ \boldsymbol{g}_0 = \boldsymbol{\Lambda}_r \boldsymbol{e}_1 \ \text{ and } \ \boldsymbol{g}_1 = \frac{\boldsymbol{r}'}{||\boldsymbol{r}'||}. \tag{2.142}$$

Accordingly, the expressions for $\boldsymbol{k}$, $\boldsymbol{w}$, $\delta\boldsymbol{\theta}$, and their respective material counterparts need to be expressed in terms of the primary fields $\boldsymbol{r}$ and φ, which is done in Sect. 2.3.3.2.

2.5.3.2 Strong and Weak Form

Identical to Sect. 2.5.2.2, the strong form defined in Eqs. (2.99) and (2.100) need to be adapted to the shear-free case. To this end, Eq. (2.100) is multiplied by $\boldsymbol{r}'$ to create the first differential equation describing the component parallel to $\boldsymbol{g}_1$ as

$$\boldsymbol{r}'^T(\boldsymbol{m}' + \boldsymbol{l} + \boldsymbol{m}_\rho + \boldsymbol{r}' \times \boldsymbol{n}) = 0. \tag{2.143}$$

In a similar manner the according perpendicular component is formulated by substracting the parallel component from the original equation resulting in

$$\begin{aligned}(\boldsymbol{m}' + \boldsymbol{l} + \boldsymbol{m}_\rho + \boldsymbol{r}' \times \boldsymbol{n}) - \boldsymbol{r}'^T(\boldsymbol{m}' + \boldsymbol{l} + \boldsymbol{m}_\rho + \boldsymbol{r}' \times \boldsymbol{n})\boldsymbol{r}' &= \\ (\boldsymbol{m}' + \boldsymbol{l} + \boldsymbol{m}_\rho)_\perp + \boldsymbol{r}' \times \boldsymbol{n}_\perp &= \boldsymbol{0}.\end{aligned} \tag{2.144}$$

Equation (2.144) can be solved for $\boldsymbol{n}_\perp$, which in turn is plugged into Eq. (2.99) to yield

$$\boldsymbol{n}'_\parallel + \left(\frac{\boldsymbol{r}'}{||\boldsymbol{r}'||^2} \times (\boldsymbol{m}' + \boldsymbol{l})\right)' + \boldsymbol{f} = 0. \tag{2.145}$$

To obtain the weak form Eqs. (2.143) and (2.145) are multiplied by $\delta\Theta_1$ (defined in Eq. (2.77)) and $\delta\boldsymbol{r}$, respectively. The integral over the whole domain is simplified by partial integrations yielding the weak form of the Kirchhoff-Love beam based on constrained rotation as

$$G_{\text{KL-CR}} = \int_0^{L_0} \left(\delta r'^T n_{\|} g_1 + \left(\delta\Theta_1 g_1 + \frac{r' \times \delta r'}{\|r'\|^2} \right)'^T m \right) \mathrm{d}s_0$$
$$- \int_0^{L_0} \left(\delta r^T (f + f_\rho) + \delta\Theta_1 g_1^T (l + m_\rho) \right) \mathrm{d}s_0 - [\delta\theta^T \overline{m} + \delta r^T \overline{n}]_{\Gamma_\sigma}. \tag{2.146}$$

Completing, proper boundary conditions

$$r = \overline{r},\ g_1 = \overline{g_1},\ \varphi = \overline{\varphi} \text{ on } \Gamma_u \text{ and } f = \overline{f},\ m = \overline{m} \text{ on } \Gamma_\sigma, \tag{2.147}$$

are introduced complementing the trial space $\mathcal{T}_{\text{KL-CR}} \subset \mathbb{R}^3 \times \mathbb{R} \ni r(s_0, t), \varphi(s_0, t)$ satisfying the above mentioned boundary conditions. $\mathcal{T}_{\text{KL-CR}}$ consists of functions with square-integrable derivatives of first and second order since the first derivative of φ and the second derivative of r are involved. Exhibiting the same properties the weighting space $\mathcal{W}_{\text{KL-CR}} \subset \mathbb{R}^3 \times \mathbb{R} \ni \delta r(s_0), \delta\varphi(s_0)$ with $\delta r = \delta\varphi = 0$ on Γ_u is specified. For further details on this approach such as removal of locking effects and comparison to other beam formulations the reader is referred to the works by Meier et al. (2014, 2015, 2017).

2.5.4 Kirchhoff-Love Beam Formulations Based on Weak Enforcement

Here, it is briefly argued how the Kirchhoff constraints from Eq. (2.115) can be enforced in a weak manner resulting in a mixed beam formulation. Namely, an additional potential due to the Lagrangian multipliers λ_2 and λ_3

$$\Pi_{\text{Lm}} = \int_0^{L_0} \left(\lambda_2 (g_2^T r') + \lambda_3 (g_3^T r') \right) \mathrm{d}s_0 \tag{2.148}$$

with the according variation

$$\delta\Pi_{\text{Lm}} = \int_0^{L_0} \left(\delta\lambda_2 (g_2^T r') + \delta\lambda_3 (g_3^T r') \right) \mathrm{d}s_0 + \int_0^{L_0} \left(\lambda_2 \delta (g_2^T r') + \lambda_3 \delta (g_3^T r') \right) \mathrm{d}s_0 \tag{2.149}$$

are introduced including a proper trial space $\mathcal{T}_{\text{Lm}} \ni \lambda_2, \lambda_3$ and weighting space $\mathcal{W}_{\text{Lm}} \ni \delta\lambda_2, \delta\lambda_3$. Some of these newly introduced quantities can be allocated physical interpretations. λ_2 and λ_3 might be interpreted as the components $g_2^T n$ and $g_3^T n$ thwarting shear deformations.

2.5.5 *Reduced Kirchhoff-Love Beam Formulations*

The most obvious possible reduction would be a restriction to isotropic and initially straight beams. Applying this simplification to the beam formulation by Meier et al. (2014) leads to a distinctive shortening of the involved equations and is identical to the formulation by Boyer & Primault (2003). But in the case of the Kirchhoff-Love beam formulation presented in Sect. 2.5.2 these restrictions do not result in a particularly simplified formulation, which is therefore not devised here.

In some applications it might be appropriate to additionally restrict the motion of the Kirchhoff-Love beam. In the case of high slenderness, axial extensibility might be neglected to further exploit the numerical benefits discussed in Sect. (2.5.1), i.e.

$$||\boldsymbol{r}'|| = \varepsilon \equiv 0 \text{ and } \dot{\varepsilon} \equiv 0 \tag{2.150}$$

is enforced everywhere along the beam. In some cases even torsional deformation can be excluded, i.e.

$$K_1 = W_1 = \dot{W} \equiv 0. \tag{2.151}$$

When modeling polymers torsional deformation does not crucially alter certain physical quantities (Cyron & Wall, 2012), which is equally true for cables of high voltage lines (Arnold & Simeon, 2000) and certain fibers of biological and artificial nature (Durville, 2010).

Enforcing these additional restrictions of vanishing axial and torsional deformations in a weak manner requires Lagrangian multipliers as done in Sect. 2.5.4 for thwarting shear deformations. This procedure is not repeated here. Instead only the strong enforcements based on the new Kirchhoff-Love beam formulation specified in Sect. 2.5.2 is detailed in the following.

2.5.5.1 Inextensible Kirchhoff-Love Beam Formulation

The strong form of the inextensible Kirchhoff-Love beam based on constraint translation is easily obtained from the strong form presented in Sect. 2.5.2.2. Simply discarding Eq. (2.126) the weak form is derived from Eq. (2.127), only, and reads

$$G_{\text{KL-IE}} = \int_0^{L_0} \left(\delta\boldsymbol{\theta}'^T\boldsymbol{m} - \delta\boldsymbol{r}^T(\boldsymbol{f} + \boldsymbol{f}_\rho) - \delta\boldsymbol{\theta}^T(\boldsymbol{l} + \boldsymbol{m}_\rho)\right) \mathrm{d}s_0 - [\delta\boldsymbol{\theta}^T\overline{\boldsymbol{m}} + \delta\boldsymbol{r}^T\overline{\boldsymbol{n}}]_{\Gamma_\sigma}. \tag{2.152}$$

The implicit definitions of $\boldsymbol{r}$ and its variations and derivatives are not repeated here. They arise by removing u' from Eqs. (2.120), (2.121), (2.122), and (2.124). Even though an application to polymer mechanics is envisioned where axial strain cannot be neglected, this element type will be closer inspected in Sect. 3.6 since it seems to be of general interest as it is the first element type that excludes shear and axial strains without the need for Lagrangian multipliers.

2.5.5.2 Torsion-free Kirchhoff-Love Beam Formulation

In contrast to the shear-free formulation, where forces perpendicular to the center line are allowed, the torsion-free formulation requires vanishing torsional external moments (Romero et al., 2014), i.e.

$$\boldsymbol{g}_1^T \boldsymbol{l} \equiv 0 \quad \text{and} \quad [\delta \boldsymbol{g}_1^T \overline{\boldsymbol{m}}]_{\Gamma_\sigma} \equiv 0. \tag{2.153}$$

This requirement leads to the fact that $\boldsymbol{l}$ is unknown beforehand and is part of the solution. Therefore, an external momentum perpendicular to $\boldsymbol{g}_1$ is prescribed as $\boldsymbol{l}_{g_1} = \boldsymbol{g}_1 \times \boldsymbol{l}$. Simply subtracting the torsional component in direction of $\boldsymbol{g}_1$ from Eq. (2.127) and including Eq. (2.126) yields the 3 differential equations to form the strong form of the torsion-free Kirchhoff-Love beam

$$\boldsymbol{g}_1^T (\boldsymbol{n}' + \boldsymbol{f} + \boldsymbol{f}_\rho) = 0 \quad \text{and} \tag{2.154}$$

$$(\boldsymbol{I} - \boldsymbol{g}_1 \otimes \boldsymbol{g}_1)\boldsymbol{m}' + \boldsymbol{l}_{g_1} \times \boldsymbol{g}_1 + \boldsymbol{m}_\rho + (1+\varepsilon)\boldsymbol{g}_1 \times \boldsymbol{n} = 0 \tag{2.155}$$

since $\boldsymbol{g}_1^T \boldsymbol{m}_\rho = 0$ and $\boldsymbol{l}_{g_1} \times \boldsymbol{g}_1 = \boldsymbol{l} - (\boldsymbol{g}_1^T \boldsymbol{l})\boldsymbol{g}_1$. All the vectors in Eq. (2.155) are elements of $T_{g_1} S^2$ and therefore, it corresponds to two scalar equations. Multiplying these relations by $\delta\varepsilon$ and $\delta\boldsymbol{\theta}_\perp$ results in the weak form

$$\begin{aligned} G_{\text{KL-TF}} = \int_0^{L_0} & \left(\delta u' n_{||} + \delta\boldsymbol{\theta}_\perp'^T \boldsymbol{m}_\perp - \delta\boldsymbol{r}^T(\boldsymbol{f} + \boldsymbol{f}_\rho) - \delta\boldsymbol{\theta}_\perp^T(\boldsymbol{l}_{g_1} \times \boldsymbol{g}_1 + \boldsymbol{m}_\rho)\right) \mathrm{d}s_0 \\ & - [\delta\boldsymbol{\theta}_\perp^T(\boldsymbol{g}_1 \times \overline{\boldsymbol{m}} \times \boldsymbol{g}_1) + \delta\boldsymbol{r}^T \overline{\boldsymbol{n}}]_{\Gamma_\sigma} \end{aligned} \tag{2.156}$$

with $\boldsymbol{m}_\perp = (\boldsymbol{I} - \boldsymbol{g}_1 \otimes \boldsymbol{g}_1)\boldsymbol{m}$. The resulting weak form constitutes a special shear-free case of the torsion-free Simo-Reissner beam presented in Romero et al. (2014). The same constitutive models from Sect. 2.5.2.3 are applicable excluding the torsional part. For a more involved discussion of the consequences of removed torsional modes on boundary conditions and constitutive relations the reader is referred to Romero et al. (2014).

To identify application fields for the torsion-free beam element Meier et al. (2017) studied, under which conditions torsion modes vanish when the simple restriction (2.153) is employed. In static cases, any torsional deformation vanishes except when both ends are clamped. In dynamics, torsion modes are negligible for highly slender structures. Further studies of torsion-free formulations are not part of this thesis since an application to polymer mechanics is envisioned where torsion does not seem to be negligible as will be seen in Sect. 4.

2.5.5.3 Weakly Removing Axial Deformation

Analogous to Sect. 2.5.4, in order to remove any axial deformation a potential due to the Lagrangian multiplier λ_ε

$$\Pi_\varepsilon = \int_0^{L_0} \lambda_\varepsilon(s_0)\varepsilon(s_0)\mathrm{d}s_0 \tag{2.157}$$

with the resulting variation

$$\delta\Pi_\varepsilon = \int_0^{L_0} (\delta\lambda_\varepsilon(s_0)\varepsilon(s_0) + \delta\varepsilon\lambda_\varepsilon)\,\mathrm{d}s_0 \tag{2.158}$$

can be implemented. Physically, λ_ε can be interpreted as axial reaction forces preventing axial deformation. It was shown by Meier et al. (2017) that the behavior of the numerical problem of weakly enforced axial inextensibility can be improved by including the contribution of axial deformation scaled by an arbitrary factor c_ε such that

$$\delta\Pi_\varepsilon = \int_0^{L_0} (\delta\lambda_\varepsilon(s_0)\varepsilon(s_0) + \delta\varepsilon(s_0)\lambda_\varepsilon(s_0) + \delta\varepsilon(s_0)c_\varepsilon EA\varepsilon(s_0))\,\mathrm{d}s_0. \tag{2.159}$$

Equation (2.159) defines an Augmented Lagrange type problem requiring proper trial and weighting spaces $\mathcal{T}_{\text{inex}} \ni \lambda_\varepsilon$ and $\mathcal{W}_{\text{inex}} \ni \delta\lambda_\varepsilon$. The factor c_ε reduces the problem arising due to the large axial stiffness EA present in Kirchhoff-Love beams, see Sect. 2.5.1.

2.5.6 *Analytical Formulation of Kirchhoff-Love Beams*

Based on the strong form defined by Eqs. (2.126) and (2.127) the Kirchhoff-Love beam is treated analytically in the following. An analytical treatment based on the alternative strong form from Eqs. (2.143) and (2.145) was obtained by Meier et al. (2015) and Drozdov & Rabin (2000) derived an analytical solution for an inextensible Kirchhoff-Love beam.

First, Eq. (2.126) is rewritten in component form relative to the local triad. To this end, it is detailed that

$$\boldsymbol{n} = [n_{||}, n_2, n_3]^T_{\boldsymbol{g}_i}, \quad \boldsymbol{f} = [f_{||}, f_2, f_3]^T_{\boldsymbol{g}_i}, \quad \text{and} \quad \boldsymbol{f}_\rho = [f_{\rho||}, f_{\rho_2}, f_{\rho_3}]^T_{\boldsymbol{g}_i}. \tag{2.160}$$

Using $\boldsymbol{g}_i' = \boldsymbol{k} \times \boldsymbol{g}_i$ from Eq. $(2.34)_1$

$$\boldsymbol{g}_1^T \boldsymbol{n}' = n_{||}' + K_2 n_3 - K_3 n_2 \tag{2.161}$$

the first differential equation of the analytical treatment can be cast as

$$n_{||}' + K_2 n_3 - K_3 n_2 + f_{||} + f_{\rho||} = 0. \tag{2.162}$$

Secondly, Eq. (2.127) is recast requiring the component form relative to the local triad of the following quantities

$$\boldsymbol{m} = [m_{||}, m_2, m_3]^T_{\boldsymbol{g}_i}, \quad \boldsymbol{l} = [l_{||}, l_2, l_3]^T_{\boldsymbol{g}_i}, \quad \text{and} \quad \boldsymbol{m}_\rho = [m_{\rho||}, m_{\rho_2}, m_{\rho_3}]^T_{\boldsymbol{g}_i}. \tag{2.163}$$

Again, using $\boldsymbol{g}_i' = \boldsymbol{k} \times \boldsymbol{g}_i$ from Eq. $(2.34)_1$ the derivative of the stress couple reads

$$\boldsymbol{m}' = (m_{||}' + K_2 m_3 - K_3 m_2)\boldsymbol{g}_1 + (m_2' - K_1 m_3 + K_3 m_1)\boldsymbol{g}_2 + (m_3' + K_1 m_2 - K_2 m_1)\boldsymbol{g}_3. \tag{2.164}$$

Projecting Eq. (2.127) into the directions of $\boldsymbol{g}_1$, $\boldsymbol{g}_2$, and $\boldsymbol{g}_3$ yields

$$\begin{aligned}
(m_{||}' + K_2 m_3 - K_3 m_2) + l_{||} + m_{\rho||} &= 0, \\
(m_2' - K_1 m_3 + K_3 m_1) + l_2 + m_{\rho 2} - (1 + \varepsilon) n_3 &= 0, \quad \text{and} \\
(m_3' + K_1 m_2 - K_2 m_1) + l_3 + m_{\rho 3} + (1 + \varepsilon) n_2 &= 0,
\end{aligned} \tag{2.165}$$

respectively. Assuming a linear constitutive model, for the stress resultants it trivially follows that

$$\begin{aligned}
n_{||} &= EA\varepsilon, \\
m_{||} &= GI_p(K_1 - K_{01}), \\
m_2 &= EI_2(K_2 - K_{02}), \quad \text{and} \\
m_3 &= EI_3(K_3 - K_{03}).
\end{aligned} \tag{2.166}$$

Plugging Eqs. (2.166) into (2.162) and (2.165) results in a system of four differential equations that can be solved for the unknown deformation measures ε, K_1, K_2, and K_3.

Chapter 3
Finite-Element Formulation of Slender Bodies Modeled by Geometrically Exact Beams

In this chapter, the beam theories are used to derive corresponding finite-element formulations. Special attention is paid to the temporal and spatial discretization of the orthogonal rotations. The beams are theoretically and numerically tested for objectivity, spatial convergence behavior, conservation properties, path-independence, and locking phenomena. In a thorough study of the performance the presented beam elements are tested and compared to beam elements from the literature.

3.1 Discretization in Time

For the temporal discretization of primary fields within the finite-element framework it is often resorted to finite difference methods. Basically, it is differentiated between explicit and implicit solution schemes. But before going into detail, some quantities are defined. First of all, in close relation to the concepts from Sect. 2.1, a motion

$$\boldsymbol{\Phi}_t : \mathcal{B} \mapsto \mathcal{V} \text{ for } t \in [0, T] \subset \mathbb{R}^+ \tag{3.1}$$

is introduced as a continuous curve of maps depending on one parameter, only, where $\mathbb{R}^+$ denotes positive rational numbers including zero. In this thesis, a constant time step size Δt is employed subdividing the total time interval $t \in [0, T]$ into equidistant sub-intervals $[t_n, t_{n+1}]$. Furthermore, an index $n \in \mathbb{N}_0$ replaces the explicit declaration of the discrete time, i.e.

$$\mathcal{B}(t_n) := \mathcal{B}_n,\ \boldsymbol{r}(s_0, t_n) := \boldsymbol{r}_n(s_0),\ \varepsilon(s_0, t_n) := \varepsilon_n(s_0) \text{ and } \boldsymbol{\Lambda}(s_0, t_n) := \boldsymbol{\Lambda}_n(s_0). \tag{3.2}$$

M. C. Schulz, *A New Kirchhoff-Love Beam Element and its Application to Polymer Mechanics*, Mechanics and Adaptronics,
https://doi.org/10.1007/978-3-031-06340-4_3

In explicit solution schemes for dynamic problems (e.g. Dokainish & Subbaraj 1989; Hulbert & Chung 1996), quantities of $\mathcal{B}_{n+1}$ are directly derived from the known configuration $\mathcal{B}_n$ resulting in a simple linear system of equations that can be solved for the unknowns in one step. For problems dominated by high frequencies or wave-like phenomena the explicit approach is favorable despite the restriction to very small values of Δt. Further simplification can be achieved by lumping techniques (e.g. Elguedj et al. 2009; Fried & Malkus 1975; Hinton et al. 1976), which are not addressed here.

In the implicit counterpart (e.g. Subbaraj & Dokainish 1989), the unknown quantities of $\mathcal{B}_{n+1}$ are involved in the problem formulation making it necessary to resort to iterative solvers. This approach is advantageous for problems dominated by low frequencies and for stiff differential equations exhibiting large differences between high and low frequencies as is the case for very slender beams. The implicit approach is of course more demanding involving additional consistent linearization.

In this thesis, known implicit time discretization schemes will be employed. Therefore, in the following, first the possible primary variables are identified. Secondly, in Sects. 3.1.2 and 3.1.3, the standard generalized-α method for elements from $\mathbb{R}^3$ and the less known counterpart for elements from $SO(3)$ are introduced in a compact manner since no new concepts are presented. Thirdly, in Sect. 3.1.4, it is commented on the choice of performing the temporal discretization prior to the spatial discretization and the choice of the primary fields. For a more comprehensive discourse on time integration techniques in mechanics the reader is referred to e.g. Hulbert (2017).

For the following considerations it is required to split the weak form, see e.g. Eqs. (2.105) and (2.128)

$$G = G_{\text{int}} + G_{\text{dyn}} - G_{\text{ext}} \tag{3.3}$$

into contributions stemming from internal, inertial, and external forces, respectively.

3.1.1 *Identification of Primary Fields*

Depending on the beam formulation at hand in spatial or material description and the orthogonal transformation interpolation strategy chosen, different primary fields can be formulated. In this section, the possible primary fields of the new beam formulations from the last section are presented. The possible time integration schemes for the Kirchhoff-Love beam based on constraint rotation presented in Sect. 2.5.3 can be found in Meier et al. (2017).

Primary fields of the Simo-Reissner beam: The standard Simo-Reissner beam presented in Sect. 2.4 can be equipped with four different rotation parameterization strategies resulting from the considerations in Sect. 2.3.3. Furthermore, based on the results by Simo & Wong (1991) the material quantities $\boldsymbol{W}$ and $\boldsymbol{A}$ are discretized rather than the spatial counterparts $\boldsymbol{w}$ and $\boldsymbol{a}$. The possible implicit time integration schemes read

$$
\begin{aligned}
(\dot{\boldsymbol{r}}, \ddot{\boldsymbol{r}}, \boldsymbol{W}, \boldsymbol{A})_{n+1} &= \boldsymbol{f}((\boldsymbol{r}, \boldsymbol{\Lambda})_{n+1}, (\boldsymbol{r}, \boldsymbol{\Lambda})_n, (\dot{\boldsymbol{r}}, \boldsymbol{W})_n, (\ddot{\boldsymbol{r}}, \boldsymbol{A})_n) \\
(\dot{\boldsymbol{r}}, \ddot{\boldsymbol{r}}, \dot{\boldsymbol{\psi}}, \ddot{\boldsymbol{\psi}})_{n+1} &= \boldsymbol{f}((\boldsymbol{r}, \boldsymbol{\Lambda})_{n+1}, (\boldsymbol{r}, \boldsymbol{\Lambda})_n, (\dot{\boldsymbol{r}}, \dot{\boldsymbol{\psi}})_n, (\ddot{\boldsymbol{r}}, \ddot{\boldsymbol{\psi}})_n) \\
(\dot{\boldsymbol{r}}, \ddot{\boldsymbol{r}}, \dot{\boldsymbol{t}}, \ddot{\boldsymbol{t}}, \dot{\varphi}, \ddot{\varphi})_{n+1} &= \boldsymbol{f}((\boldsymbol{r}, \boldsymbol{\Lambda})_{n+1}, (\boldsymbol{r}, \boldsymbol{\Lambda})_n, (\dot{\boldsymbol{r}}, \dot{\boldsymbol{t}}, \dot{\varphi})_n, (\ddot{\boldsymbol{r}}, \ddot{\boldsymbol{t}}, \ddot{\varphi})_n) \\
(\dot{\boldsymbol{r}}, \ddot{\boldsymbol{r}}, \dot{\boldsymbol{t}}, \ddot{\boldsymbol{t}}, W_1, \dot{W}_1)_{n+1} &= \boldsymbol{f}((\boldsymbol{r}, \boldsymbol{\Lambda})_{n+1}, (\boldsymbol{r}, \boldsymbol{\Lambda})_n, (\dot{\boldsymbol{r}}, \dot{\boldsymbol{t}}, W_1)_n, (\ddot{\boldsymbol{r}}, \ddot{\boldsymbol{\psi}}, \dot{W}_1)_n).
\end{aligned} \tag{3.4}
$$

The first scheme appears to be the most favorable one as it could be applied in a simple manner to the Simo-Reissner beam presented in Sect. 2.4. The primary fields defined in Eq. (3.4).2 would require more complicated formulations, see Eq. (2.45). Approaches (3.4).3 and (3.4).4 seem to be unfavorable for Simo-Reissner beams since one additional degree of freedom is required. Despite these apparent drawbacks, a comparative study of different implementations might be an interesting endeavor.

Primary fields of the Kirchhoff-Love beam based on constraint translation: Different choices of primary fields for the new Kirchhoff-Love beam detailed in Sect. 2.5.2 are introduced. Four different approaches are available building on the kinematic considerations of Sect. 2.3.3. As mentioned before, following Simo & Wong (1991) $\boldsymbol{W}$ and $\boldsymbol{A}$ are discretized rather than $\boldsymbol{w}$ and $\boldsymbol{a}$.

$$
\begin{aligned}
(\dot{\varepsilon}, \ddot{\varepsilon}, \dot{\boldsymbol{r}}^*, \ddot{\boldsymbol{r}}^*, \boldsymbol{W}, \boldsymbol{A})_{n+1} &= \boldsymbol{f}((\varepsilon, \boldsymbol{r}^*, \boldsymbol{\Lambda})_{n+1}, (\varepsilon, \boldsymbol{r}^*, \boldsymbol{\Lambda})_n, (\dot{\varepsilon}, \dot{\boldsymbol{r}}^*, \boldsymbol{W})_n, (\ddot{\varepsilon}, \ddot{\boldsymbol{r}}^*, \boldsymbol{A})_n) \\
(\dot{\varepsilon}, \ddot{\varepsilon}, \dot{\boldsymbol{r}}^*, \ddot{\boldsymbol{r}}^*, \dot{\boldsymbol{\psi}}, \ddot{\boldsymbol{\psi}})_{n+1} &= \boldsymbol{f}((\varepsilon, \boldsymbol{r}^*, \boldsymbol{\Lambda})_{n+1}, (\varepsilon, \boldsymbol{r}^*, \boldsymbol{\Lambda})_n, (\dot{\varepsilon}, \dot{\boldsymbol{r}}^*, \dot{\boldsymbol{\psi}})_n, (\ddot{\varepsilon}, \ddot{\boldsymbol{r}}^*, \ddot{\boldsymbol{\psi}})_n) \\
(\dot{\boldsymbol{r}}^*, \ddot{\boldsymbol{r}}^*, \dot{\boldsymbol{t}}, \ddot{\boldsymbol{t}}, \dot{\varphi}, \ddot{\varphi})_{n+1} &= \boldsymbol{f}((\boldsymbol{r}^*, \boldsymbol{t}, \boldsymbol{\Lambda})_{n+1}, (\boldsymbol{r}^*, \boldsymbol{t}, \boldsymbol{\Lambda})_n, (\dot{\varphi}, \dot{\boldsymbol{r}}^*, \dot{\boldsymbol{t}})_n, (\ddot{\varphi}, \ddot{\boldsymbol{r}}^*, \ddot{\boldsymbol{t}})_n) \\
(\dot{\boldsymbol{r}}^*, \ddot{\boldsymbol{r}}^*, \dot{\boldsymbol{t}}, \ddot{\boldsymbol{t}}, W_1, \dot{W}_1)_{n+1} &= \boldsymbol{f}((\boldsymbol{r}^*, \boldsymbol{t}, \boldsymbol{\Lambda})_{n+1}, (\boldsymbol{r}^*, \boldsymbol{t}, \boldsymbol{\Lambda})_n, (W_1, \dot{\boldsymbol{r}}^*, \dot{\boldsymbol{t}})_n, (\dot{W}_1, \ddot{\boldsymbol{r}}^*, \ddot{\boldsymbol{t}})_n).
\end{aligned} \tag{3.5}
$$

Again, as was the case when studying the primary field of the Simo-Reissner beam, variants (3.5).2–(3.5).4 are distinctively more complicated but could be studied in more detail in future works. In the scope of this thesis, variant (3.5).1 is implemented, only, mainly due to its compactness and flexibility.

3.1.2 Generalized-α Method for Elements of $\mathbb{R}$ and $\mathbb{R}^3$

In structural and solid mechanics, the Newmark time integration method (Newmark, 1959) is one of the most popular methods (Borst et al., 2012; Wriggers, 2008). Here, the concept is introduced representatively for $\boldsymbol{r}$ and its derivatives. The discretization of other primary fields such as ε, $\boldsymbol{t}$, and W_1 is not explicitly given. In the Newmark scheme, it is assumed that the acceleration varies linearly over the time step such that

$$
\begin{aligned}
\boldsymbol{r}_{n+1} &= \boldsymbol{r}_n + \Delta t \dot{\boldsymbol{r}}_n + \Delta t^2[(0.5 - \beta)\ddot{\boldsymbol{r}}_n + \beta \ddot{\boldsymbol{r}}_{n+1}] \\
\dot{\boldsymbol{r}}_{n+1} &= \dot{\boldsymbol{r}}_n + \Delta t[(1 - \gamma)\ddot{\boldsymbol{r}}_n + \gamma \ddot{\boldsymbol{r}}_{n+1}].
\end{aligned} \tag{3.6}
$$

It was shown for the linear case that the integration parameters are limited to $\beta \in [0, 0.5]$ and $\gamma \in [0, 1]$ (Bathe, 1982). The particular choice of β and γ, made by the modeler, determines the stability, accuracy, dissipation, and dispersion of the integration scheme. An overview of the Newmark variants of time integration and their respective properties can be found in Cook (2007). For instance, setting $\beta = 0$ leads to the explicit Newmark scheme, which is excluded here. For future reference, the approximations (3.6) are reformulated to

$$\begin{aligned}\dot{\boldsymbol{r}}_{n+1} &= \frac{\gamma}{\beta \Delta t}(\boldsymbol{r}_{n+1} - \boldsymbol{r}_n) + \left(1 - \frac{\gamma}{\beta}\right)\dot{\boldsymbol{r}}_n + \left(1 - \frac{\gamma}{2\beta}\right)\Delta t \ddot{\boldsymbol{r}}_n \\ \ddot{\boldsymbol{r}}_{n+1} &= \frac{1}{\beta \Delta t^2}(\boldsymbol{r}_{n+1} - \boldsymbol{r}_n) - \frac{1}{\beta \Delta t}\dot{\boldsymbol{r}}_n + \left(1 - \frac{1}{2\beta}\right)\ddot{\boldsymbol{r}}_n.\end{aligned} \tag{3.7}$$

The Newmark approach reveals one of its weaknesses when numerical dissipation needs to be introduced to reduce the extent of high frequency modes caused by spatial discretization (Wriggers, 2008). Also, Lagrangian multipliers can cause numerical oscillations, which can be treated by numerical damping (Cardona & Geradin, 1989). In order to achieve that, $\gamma > 0.5$ would be required but only for $\gamma = 0.5$ the Newmark method produces second-order accuracy (Wood, 1990). To remedy this deficiency the generalized-α method has been proposed by Chung & Hulbert (1993), of which the Newmark time integration scheme, the HHT-α method (Hilber et al., 1977), and the WBZ-α method (Wood et al., 1980) are special cases.

Following Chung & Hulbert (1993), two new algorithmic parameters $\alpha_f \in [0, 1]$ and $\alpha_m \in [0, 1]$ are introduced defining two midpoints within a time interval. At these midpoints the kinematic quantities

$$\begin{aligned}\boldsymbol{r}_{\text{mod},n+1} &= (1 - \alpha_f)\boldsymbol{r}_{n+1} + \alpha_f \boldsymbol{r}_n \\ \dot{\boldsymbol{r}}_{\text{mod},n+1} &= (1 - \alpha_f)\dot{\boldsymbol{r}}_{n+1} + \alpha_f \dot{\boldsymbol{r}}_n \\ \ddot{\boldsymbol{r}}_{\text{mod},n+1} &= (1 - \alpha_m)\ddot{\boldsymbol{r}}_{n+1} + \alpha_m \ddot{\boldsymbol{r}}_n \;\text{ with }\; \ddot{\boldsymbol{r}}_{\text{mod}}(t = 0) = \ddot{\boldsymbol{r}}_0\end{aligned} \tag{3.8}$$

and contributions to the weak form

$$\begin{aligned}G_{\text{int, mod},n+1} &= (1 - \alpha_f)G_{\text{int},n+1} + \alpha_f G_{\text{int},n} \\ G_{\text{ext, mod},n+1} &= (1 - \alpha_f)G_{\text{ext},n+1} + \alpha_f G_{\text{ext},n}\end{aligned} \tag{3.9}$$

are evaluated using a simple interpolation according to the trapezoidal rule. There are also other interpolation strategies such as the midpoint rule. The last relation of Eq. (3.8) is updated upon convergence and does not contribute to the assembled linearizations forming the stiffness matrix. Furthermore, the real and modified accelerations are related by

$$(1 - \alpha_m)\ddot{\boldsymbol{r}}_{\text{mod},n+1} + \alpha_m \ddot{\boldsymbol{r}}_{\text{mod},n} = (1 - \alpha_f)\ddot{\boldsymbol{r}}_{n+1} + \alpha_f \ddot{\boldsymbol{r}}_n. \tag{3.10}$$

According to these newly introduced quantities, the Newmark formulation specified by Eq. (3.7) are slightly reformulated yielding

$$
\begin{aligned}
\Delta\tilde{\boldsymbol{r}}_{n+1} &:= \boldsymbol{r}_{n+1} - \boldsymbol{r}_n, \\
\dot{\boldsymbol{r}}_{n+1} &= \frac{\gamma}{\beta\Delta t}\Delta\tilde{\boldsymbol{r}}_{n+1} + \left(1 - \frac{\gamma}{\beta}\right)\dot{\boldsymbol{r}}_n + \Delta t\left(1 - \frac{\gamma}{2\beta}\right)\ddot{\boldsymbol{r}}_{\mathrm{mod},n}, \\
\ddot{\boldsymbol{r}}_{n+1} &= \frac{1-\alpha_m}{\beta\Delta t^2(1-\alpha_f)}\Delta\tilde{\boldsymbol{r}}_{n+1} - \frac{1-\alpha_m}{\beta\Delta t(1-\alpha_f)}\dot{\boldsymbol{r}}_n \\
&\quad + \left(-\frac{(1-\alpha_m)(0.5-\beta)}{\beta(1-\alpha_f)} + \frac{\alpha_m}{1-\alpha_f}\right)\ddot{\boldsymbol{r}}_{\mathrm{mod},n} - \frac{\alpha_f}{1-\alpha_f}\ddot{\boldsymbol{r}}_n.
\end{aligned} \tag{3.11}
$$

A tilde is added for $\Delta\tilde{\boldsymbol{r}}$ to stress that the increment is in relation to the last converged time step and not in relation to the last Newton iteration. To achieve second-order accuracy, unconditional stability in the linear regime, and favorable damping of the high- and low-frequencies, the numerical parameters are constrained to a single user-defined parameter, the so-called spectral radius $\rho_\infty \in [0, 1]$, see e.g. Erlicher et al. (2002), such that

$$
\alpha_m = \frac{2\rho_\infty - 1}{\rho_\infty + 1}, \quad \alpha_f = \frac{\rho_\infty}{\rho_\infty + 1}, \quad \beta = \frac{(1-\alpha_m+\alpha_f)^2}{4}, \quad \gamma = \frac{1}{2} - \alpha_m + \alpha_f. \tag{3.12}
$$

For the choice $\rho_\infty = 0$, the high-frequency response is annihilated asymptotically and for $\rho_\infty = 1$ no algorithmic dissipation occurs. Furthermore, the Newmark family is obtained when $\alpha_f = 0, \alpha_m = 0$, the HHT-$\alpha$ method is retrieved for $\alpha_f \neq 0, \alpha_m = 0$, and the WBZ-$\alpha$ scheme is regained for $\alpha_f = 0$, $\alpha_m \neq 0$. For an in-depth analysis of the generalized-α method the reader is referred to e.g. Arnold & Brüls (2007), Baldo (2006), and Erlicher et al. (2002). Therein, the time discretization is applied to the spatially discretized equations, which is the more common approach. Furthermore, it is crucial to note that the time integration method presented so far were actually developed for linear problems. By implication this means that applying these methods to nonlinear problems might lead to non-preservation of physical quantities such as linear momentum, angular momentum and mechanical energy. The possible presence of such problems will be studied in a numerical example in Sect. 3.6.

Summarizing, in the present section, Eq. (3.3) without rotational degrees of freedom has been transformed to a semi-discrete variant

$$
G_{\mathrm{int,mod}} + G_{\mathrm{dyn}}(\ddot{\boldsymbol{r}}_{\mathrm{mod}}) - G_{\mathrm{ext,mod}} \doteq 0, \tag{3.13}
$$

which further requires spatial discretization.

3.1.3 Generalized-α Method for Elements of SO(3)

As mentioned in the preceding Sects. 3.1.1 and 3.1.1, the time integration scheme is applied to the material quantities $\boldsymbol{W}$ and $\boldsymbol{A}$, which are elements of tangent spaces to $SO(3)$. The standard generalized-α method has been extended to Lie-Groups by Arnold & Brüls (2007), Brüls & Cardona (2010), and Brüls et al. (2012). Interestingly, the authors found that the same parameter choice as stated in Eq. (3.12) leads to second-order accuracy, unconditional stability in the linear regime, and favorable damping. Furthermore, the scheme presented in this section can consistently include non-constant mass matrix contributions and constraint systems involving e.g. Lagrangian multipliers. Before continuing, different incremental rotation vectors are defined. An iterative Newton-Raphson solution procedure with iterations denoted by the subscript k is used. Incremental material and spatial rotations between the last converged and the current, possibly unconverged time step are denoted by $\tilde{\boldsymbol{\Theta}}$ and $\tilde{\boldsymbol{\theta}}$, respectively. The material and spatial incremental rotations between iterations within a time step are denoted by $\Delta\boldsymbol{\Theta}$ and $\Delta\boldsymbol{\theta}$, respectively. These increments are implicitly defined as

$$\boldsymbol{\Lambda}_{n+1,k} = \boldsymbol{\Lambda}_n \exp(\boldsymbol{S}(\tilde{\boldsymbol{\Theta}}_{n+1,k})) = \exp(\boldsymbol{S}(\tilde{\boldsymbol{\theta}}_{n+1,k}))\boldsymbol{\Lambda}_n \tag{3.14}$$

and

$$\boldsymbol{\Lambda}_{n,k+1} = \boldsymbol{\Lambda}_{n,k} \exp(\boldsymbol{S}(\Delta\boldsymbol{\Theta}_{k+1})) = \exp(\boldsymbol{S}(\Delta\boldsymbol{\theta}_{k+1}))\boldsymbol{\Lambda}_{n,k}, \tag{3.15}$$

respectively.

The material and spatial increments are related by $\tilde{\boldsymbol{\Theta}}_{n+1} = \boldsymbol{\Lambda}_n^T \tilde{\boldsymbol{\theta}}_{n+1} = \boldsymbol{\Lambda}_{n+1}^T \tilde{\boldsymbol{\theta}}_{n+1}$, which holds equally for $\Delta\tilde{\boldsymbol{\Theta}}$ and $\Delta\tilde{\boldsymbol{\theta}}$. An increment of the rotation described by $\tilde{\boldsymbol{\theta}}$ or $\tilde{\boldsymbol{\Theta}}$ can be related to an increment of the rotation described by $\boldsymbol{\theta}$ via

$$\Delta\tilde{\boldsymbol{\Theta}}_{n+1,k} = \boldsymbol{\Lambda}_n^T \Delta\tilde{\boldsymbol{\theta}}_{n+1,k} = \boldsymbol{\Lambda}_n^T \boldsymbol{T}(\tilde{\boldsymbol{\theta}}_{n+1,k})\Delta\boldsymbol{\theta}_{n+1,k}, \tag{3.16}$$

with (Jelenić & Crisfield, 1999)

$$\boldsymbol{T}(\tilde{\boldsymbol{\theta}}) = \frac{1}{\tilde{\theta}}\tilde{\boldsymbol{\theta}} \otimes \tilde{\boldsymbol{\theta}} + \frac{\tilde{\theta}/2}{\tan(\tilde{\theta}/2)}(\boldsymbol{I} - \frac{1}{\tilde{\theta}^2}\tilde{\boldsymbol{\theta}} \otimes \tilde{\boldsymbol{\theta}}) - \frac{1}{2}\boldsymbol{S}(\tilde{\boldsymbol{\theta}}). \tag{3.17}$$

The generalized-α method derived for elements of $SO(3)$ then reads

$$\begin{aligned} \boldsymbol{W}_{n+1} &= \frac{\gamma}{\beta\Delta t}\tilde{\boldsymbol{\Theta}}_{n+1} + \left(1 - \frac{\gamma}{\beta}\right)\boldsymbol{W}_n + \Delta t\left(1 - \frac{\gamma}{2\beta}\right)\boldsymbol{A}_{mod,n}, \\ \boldsymbol{A}_{n+1} &= \frac{1-\alpha_m}{\beta\Delta t^2(1-\alpha_f)}\tilde{\boldsymbol{\Theta}}_{n+1} - \frac{1-\alpha_m}{\beta\Delta t(1-\alpha_f)})\boldsymbol{W}_n \\ &\quad + \left(-\frac{(1-\alpha_m)(0.5-\beta)}{\beta(1-\alpha_f)} + \frac{\alpha_m}{1-\alpha_f}\right)\boldsymbol{A}_{mod,n} - \frac{\alpha_f}{1-\alpha_f}\boldsymbol{A}_n, \end{aligned} \tag{3.18}$$

where the relation

$$\boldsymbol{A}_{mod,n+1} = \frac{1}{1-\alpha_m}\big((1-\alpha_f)\boldsymbol{A}_{n+1} + \alpha_f\boldsymbol{A}_n - \alpha_m\boldsymbol{A}_{mod,n}\big) \quad \text{with} \quad \boldsymbol{A}_{mod,0} = \boldsymbol{A}_0 \tag{3.19}$$

is used. When $\alpha_f = 0$, $\alpha_m = 0$ is chosen, the presented integration scheme reduces to the variant presented by Simo & Vu-Quoc (1988). It should be noted that the conserved physical quantities need to be monitored as a conservation for nonlinear problems is not guaranteed. The presented method is chosen due to its flexibility and simplicity as compared to other strategies, which aim to conserve energy, linear and angular momentum (Bauchau & Bottasso, 1999; Betsch & Steinmann, 2002; Bottasso & Borri, 1998; Demoures et al., 2015; Gonzalez, 1996; Ibrahimbegovic & Mamouri, 2002; Kane et al., 2000; Lens & Cardona, 2008; Romero, 2008b; Simo & Wong, 1991). It can be easily applied to Simo-Reissner and Kirchhoff-Love beams.

Concluding, in the present section, Eq. (3.3) including both translational and rotational degrees of freedom has been transformed to a semi-discrete variant

$$G_{\text{int,mod}} + G_{\text{dyn}}(\ddot{\boldsymbol{r}}_{\text{mod}}, \boldsymbol{A}_{\text{mod}}) - G_{\text{ext,mod}} \doteq 0, \tag{3.20}$$

which still requires spatial discretization.

3.1.4 Order of Temporal and Spatial Discretization

In this section, following Meier et al. (2017), it is briefly presented that the order of temporal and spatial discretization of the weak form alters the resulting equations. Moreover, it becomes obvious why it is clearly favorable to start by a temporal discretization as usually done with beam formulations.

The orthogonal rotations are parameterized by rotation vectors according to Sect. 2.3.3 and the time integration is realized using the Lie group extension of the generalized-α Newmark scheme presented in Sect. 3.1.3. Without loss of generality it is assumed that the rotation within an element is spatially discretized by a rotation vector

$$\boldsymbol{\psi}_h(\xi) = \mathbf{f}(\boldsymbol{\psi}^1, \boldsymbol{\psi}^2), \tag{3.21}$$

where $\boldsymbol{\psi}^1$ and $\boldsymbol{\psi}^2$ are the nodal rotation vectors and ξ denotes the local coordinate. Numbers in the superscript indicate an affiliation of the quantity with the respective node within the finite-element scheme. Furthermore, Eq. (2.45) is required to relate $\boldsymbol{W}$ to $\dot{\boldsymbol{\psi}}$.

In the first case studied, $\boldsymbol{W}$ is temporally and subsequently spatially discretized yielding in condensed form

$$\boldsymbol{W}_{n+1}(s) = \frac{\gamma}{\beta \Delta t} \tilde{\boldsymbol{\Theta}}_{n+1} + [...] \text{ and}$$

$$\boldsymbol{W}_{n+1}(\xi) = \frac{\gamma}{\beta \Delta t} \left(\text{sp} \left[\exp \left\{ \boldsymbol{S}(\mathbf{f}(\boldsymbol{\psi}^1_{n+1}, \boldsymbol{\psi}^2_{n+1}, \xi)) \right\}^T \exp \left\{ \boldsymbol{S}(\mathbf{f}(\boldsymbol{\psi}^1_n, \boldsymbol{\psi}^2_n, \xi)) \right\} \right] \right) + [...] \,. \tag{3.22}$$

Conversely, starting with the spatial discretization, the following shortened relations are obtained

$$
\begin{aligned}
\boldsymbol{W}(s,t) &= \boldsymbol{\Lambda}^T \boldsymbol{T}^{-1}(\boldsymbol{\psi})\dot{\boldsymbol{\psi}}, \\
\boldsymbol{W}(\xi,t) &= \boldsymbol{\Lambda}^T \boldsymbol{T}^{-1}(\boldsymbol{\psi}_h)\left(\frac{\partial \mathbf{f}}{\partial \boldsymbol{\psi}^1}\boldsymbol{T}(\boldsymbol{\psi}^1)\boldsymbol{\Lambda}^1\boldsymbol{W}^1 + \frac{\partial \mathbf{f}}{\partial \boldsymbol{\psi}^2}\boldsymbol{T}(\boldsymbol{\psi}^2)\boldsymbol{\Lambda}^2\boldsymbol{W}^2\right), \\
\boldsymbol{W}_{n+1}(\xi) &= \boldsymbol{\Lambda}^T \boldsymbol{T}^{-1}(\boldsymbol{\psi}_h) \\
&\quad \times \left(\frac{\partial \mathbf{f}}{\partial \boldsymbol{\psi}^1}\boldsymbol{T}(\boldsymbol{\psi}^1)\boldsymbol{\Lambda}^1\frac{\gamma}{\beta\Delta t}\left(\mathrm{sp}\left[\exp\left\{\boldsymbol{S}(\boldsymbol{\psi}^1_{n+1})\right\}^T \exp\left\{\boldsymbol{S}(\boldsymbol{\psi}^1_n)\right\}\right]\right)\right. \\
&\quad \left. + \frac{\partial \mathbf{f}}{\partial \boldsymbol{\psi}^2}\boldsymbol{T}(\boldsymbol{\psi}^2)\boldsymbol{\Lambda}^2\frac{\gamma}{\beta\Delta t}\left(\mathrm{sp}\left[\exp\left\{\boldsymbol{S}(\boldsymbol{\psi}^2_{n+1})\right\}^T \exp\left\{\boldsymbol{S}(\boldsymbol{\psi}^2_n)\right\}\right]\right)\right).
\end{aligned}
\tag{3.23}
$$

The variant presented in Eq. (3.23) is computationally more demanding than the one in (3.22) since it involves the inverse $\boldsymbol{T}^{-1}$ and its temporal derivative, see Eq. (2.45). Thus, in this thesis, as in many other works, the weak form is discretized first temporally and then spatially.

3.2 Discretization in Space

This section starts with a brief presentation of the finite-element method as far as required to introduce the notation. For a more involved treatise on the finite-element method the reader is referred to standard textbooks, e.g. Bathe (2006), Belytschko et al. (2013), Brezzi & Fortin (1991), Hughes (2012), Reddy (2013), Reddy (2014), Strang & Fix (2008), Zienkiewicz & Taylor (2005), and Zienkiewicz et al. (2005).

For the sake of conciseness the dependency on the time t is dropped in this section, which is concerned exclusively with spatial discretization.

3.2.1 Essentials of the Finite-Element Method

The weak forms defined so far for the Simo-Reissner beam in Eq. (2.105), for the Kirchhoff-Love beam based on constraint translation in Eq. (2.128), for the Kirchhoff-Love beam based on constraint rotation in Eq. (2.146), for reduced Kirchhoff-Love beam formulations in Eq. (2.152) as well as (2.156), and for the weak formulations based on weak enforcement of the Kirchhoff constraint defined in Eqs. (2.149) and (2.159) are all solved using the finite-element method. To briefly and conveniently introduce the method, weak forms G, primary fields or test functions $\boldsymbol{q}$, virtual primary fields or weighting functions $\delta\boldsymbol{q}$, the test function space $\mathcal{T}$, and the weighting function space $\mathcal{W}$ are referred to in the following. In this notation, the task at hand is:

$$\text{Find}(\boldsymbol{q}) \in \mathcal{T} \quad \text{such that} \quad G(\boldsymbol{q}, \delta\boldsymbol{q}) = 0 \quad \forall\, \delta\boldsymbol{q} \in \mathcal{W}. \tag{3.24}$$

It is assumed that this problem has a unique solution. Then, the Galerkin approach lies at the core of the finite-element method where the analytic and continuous fields $\boldsymbol{q}$ and $\delta\boldsymbol{q}$ are approximated by discrete representations $\boldsymbol{q}_h$ and $\delta\boldsymbol{q}_h$. In general, a discrete representation is obtained by a sum of weighted so-called shape functions from an appropriate function space. In the finite-element method, the shape functions are defined locally within finite elements, which results in considerable benefits for the numerical procedure. Thus, the domain of the beam $\mathcal{B}$ is discretized into a total number of n_{ele} sub-domains (elements) such that $\mathcal{B} \approx \mathcal{B}_h = \bigcup_{e=1}^{n_{\text{ele}}} \mathcal{B}_h^e$. Each element consist of at least two boundary nodes with an arbitrary amount of additional nodes within the element yielding n_{node} nodes per element in total. The discrete values of $\boldsymbol{q}_h$ and $\delta\boldsymbol{q}_h$ at the node i in the element e are designated $\hat{\boldsymbol{q}}_i^e$ and $\delta\hat{\boldsymbol{q}}_i^e$, respectively, for $i = 1, ..., n_{\text{node}}$. Summarizing, the process of discretization constitutes an introduction of confined finite-dimensional trial and weighting sub-spaces $\mathcal{T}_h \subset \mathcal{T}$ and $\mathcal{W}_h \subset \mathcal{W}$, respectively. The discrete counterpart of (3.24) can now be formulated as

$$\text{Find}(\boldsymbol{q}_h) \in \mathcal{T}_h \quad \text{such that} \quad G(\boldsymbol{q}_h, \delta\boldsymbol{q}_h) = 0 \quad \forall\, \delta\boldsymbol{q}_h \in \mathcal{W}_h. \tag{3.25}$$

To further simplify the numerical scheme, the physical space of each element is mapped to a parent space $\xi \in [-1, +1]$. The mapping is realized by the element Jacobian

$$J^e(\xi) := \left\| \frac{\mathrm{d}\boldsymbol{r}_{0h}^e}{\mathrm{d}\xi} \right\| \quad \text{such that} \quad \mathrm{d}s_0^e = J^e(\xi)\mathrm{d}\xi, \tag{3.26}$$

where the superscript $(\bullet)^e$ tags a local quantity considered within en element. J^e obviously depends on the discretization of $\boldsymbol{r}_0$. The change of parameter space brings the benefit that the shape functions can be expressed in terms of ξ and are no longer dependent on a specific element. In Sect. 3.2.2 it is commented on the computation of J^e. Furthermore, the isoparametric approach is chosen where the undeformed and deformed geometry and their relation are expressed using the same shape functions. When the same shape functions are used to discretize the trial and weighting functions a Bubnov-Galerkin approach is followed. The opposite choice is referred to as Petrov-Galerkin approach.

Finally, the two terms *consistency* and *stability* are briefly touched on. Let m denote the highest occurring derivative in G and p the order of the employed trial functions. The finite-element formulation is said to be consistent when $\boldsymbol{q}_h$ can exactly represent polynomials of order p and when in addition derivatives of $\boldsymbol{q}_h$ of order m are square-integrable. These conditions are tightly related to the continuity requirements of $\boldsymbol{q}_h$ at the element boundaries and the boundedness of G_h. Typically, C^{m-1}-continuous shape functions are employed to ensure the second requirement is met, which is also done here.

The discretized formulation is considered stable when it fulfills certain conditions, e.g. the presence of a theoretically expected convergence order and the absence of non-physical zero-energy modes and locking effects. The latter is studied in Sect. 3.5.3. For further details the references listed above can be consulted.

In the remainder of this work, the subscript $(\bullet)_h$ marks a spatially discretized quantity, the superscript $(\bullet)^e$ tags a local quantity considered within en element, and a hat $(\hat{\bullet})$ indicates a nodal quantity.

Discrete problem formulation: To conveniently present the remaining topics of the finite-element method some new quantities are introduced, which will be particularized in the following sections where the different elements are presented. The nodal values of the test and weighting functions used to weight the shape functions within an element are summarized as

$$\hat{\mathbf{q}}^e := \left(\hat{\boldsymbol{q}}_1^{eT}, ..., \hat{\boldsymbol{q}}_{n_\text{node}}^{eT}\right)^T \text{ and } \delta\hat{\mathbf{q}}^e := \left(\delta\hat{\boldsymbol{q}}_1^{eT}, ..., \delta\hat{\boldsymbol{q}}_{n_\text{node}}^{eT}\right)^T. \tag{3.27}$$

The collections of nodal values $\hat{\boldsymbol{q}}_i^e$ and $\delta\hat{\boldsymbol{q}}_i^e$ need not be of equal dimension for $i = 1, ..., n_\text{node}$. For example, in the Kirchhoff-Love beam formulation based on constrained rotation the interpolation of the orthogonal rotation requires three nodes while for the interpolation of the center line only two nodes are employed, see Meier et al. (2014). Furthermore, the element residual vector $\hat{\mathbf{r}}^e$ is defined such that

$$G_h^e := \delta\hat{\mathbf{q}}^{eT}\mathbf{r}^e \text{ with } \mathbf{r}^e := \int_{s_0^i}^{s_0^{i+1}} [...] \, \mathrm{d}s_0. \tag{3.28}$$

These quantities defined on element level need to be assembled to obtain the corresponding global expressions

$$\mathbf{Q} := \mathop{\mathsf{A}}_{e=1}^{n_\text{ele}} \hat{\mathbf{q}}^e, \quad \delta\mathbf{Q} := \mathop{\mathsf{A}}_{e=1}^{n_\text{ele}} \delta\hat{\mathbf{q}}^e, \text{ and } \mathbf{R} := \mathop{\mathsf{A}}_{e=1}^{n_\text{ele}} \mathbf{r}^e, \tag{3.29}$$

where $\mathbf{Q}$ defines the global collection of the nodal values of the primary fields, $\delta\mathbf{Q}$ contains the corresponding values of the weighting functions, and $\mathbf{R}$ is the global residual vector. Furthermore, Asignifies the typical procedure of assembling the contributions of the local element nodes over n_ele sub-domains to global quantities. Applying a generalized-α time integration scheme as specified in Eqs. (3.11) and (3.18), at time t_n the global residual vector depends on $\mathbf{Q}_n$, only, such that the total discretized problem can be reformulated as

$$G_h = \delta\mathbf{Q}^T\mathbf{R}(\mathbf{Q}_n) \doteq 0. \tag{3.30}$$

Usually, $\mathbf{R}$ depends non-linearly on $\mathbf{Q}_n$ requiring linearization in a Newton-Raphson type iterative solution scheme, which is treated in the next section.

Linearization of global residual vector and characteristics of global stiffness matrix: The discrete problem from Eq. (3.30) requires linearization to be solved in an iterative way. To this end, the element and global stiffness matrices

$$\mathbf{k}^e := \frac{\mathrm{d}\mathbf{r}^e}{\mathrm{d}\hat{\mathbf{q}}^e} \quad \text{and} \quad \mathbf{K} := \mathop{\mathsf{A}}_{e=1}^{n_{\mathrm{ele}}} \mathbf{k}^e, \tag{3.31}$$

respectively, are defined to be used as the first term in a truncated Taylor series

$$\mathrm{L}(\delta\mathbf{Q}^T\mathbf{R})\,\big|_{\mathbf{Q}_n} := \delta\mathbf{Q}^T\,(\mathbf{R}(\mathbf{Q}_n) + \mathbf{K}(\mathbf{Q}_n)\Delta\mathbf{Q}_n) \doteq 0. \tag{3.32}$$

Therein, $\delta\mathbf{Q}$ can be simply disregarded and $\mathrm{L}(\bullet)\,\big|_{\mathbf{Q}_n}$ designates the linearization operator to be evaluated at $\mathbf{Q}_n$.

To concisely point out some well-known special properties of the stiffness matrices appearing in finite-element formulations involving rotational degrees of freedom (see e.g. Argyris et al. (1979), Argyris (1982), Simo (1985), and Ziegler (2013)) the element stiffness and residual matrices are split into contributions stemming from internal forces $(\bullet)_{\mathrm{int}}$, from inertial forces $(\bullet)_{\mathrm{kin}}$, and from external forces $(\bullet)_{\mathrm{ext}}$ such that

$$\mathbf{k}^e = \mathbf{k}^e_{\mathrm{int}} + \mathbf{k}^e_{\mathrm{kin}} - \mathbf{k}^e_{\mathrm{ext}} := \frac{\mathrm{d}\mathbf{r}_{\mathrm{int}}}{\mathrm{d}\hat{\mathbf{q}}^e} + \frac{\mathrm{d}\mathbf{r}_{\mathrm{kin}}}{\mathrm{d}\hat{\mathbf{q}}^e} - \frac{\mathrm{d}\mathbf{r}_{\mathrm{ext}}}{\mathrm{d}\hat{\mathbf{q}}^e}. \tag{3.33}$$

Symmetry is a desired property for matrices appearing in systems of linear equations since this property can be exploited to greatly reduce the computational effort. However, due to the non-conservative external axial moments, which were already discussed in Sect. 2.5.2.3, and due to the involved non-linear configuration space requiring multiplicative update procedures, see Eqs. (2.40) and (2.88), non-symmetric stiffness matrices are found. This finding holds for both, the continuous and the discretized problem.

In general, $\mathbf{k}^e_{\mathrm{int}}$ is symmetric only for an equilibrium configuration when in addition no axial moments are applied or for reformulated beam models such as reduced models where no non-linear configuration space is found, see e.g. the torsion-free formulation in Meier et al. (2017). Analogously, $\mathbf{k}^e_{\mathrm{kin}}$ turns out to be generally non-symmetric due to the involved rotational degrees of freedom. Non-vanishing $\mathbf{k}^e_{\mathrm{ext}}$ occur due to deformation-dependent loads (follower loads) and due to deformation-dependent weighting functions such as $\delta\boldsymbol{r}$ and $\delta\boldsymbol{\theta}$ in Kirchhoff-Love beams based on constraint translation and rotation, respectively. Weighting functions with non-vanishing first derivatives with respect to the primary variables do also occur in the Bubnov-Galerkin variants of rotation interpolation schemes, see e.g. Jelenić & Crisfield (1999).

Solution schemes for linearized and discretized problems: To solve the discretized problem formulated in Eq. (3.25) a classical Newton-Raphson scheme is applied using the linearized problem formulation given in Eq. (3.32). In detail, at each time step t_{n+1}

$$\mathbf{K}\left(\mathbf{Q}_n^k\right)\Delta\mathbf{Q}_{n+1}^{k+1} = -\mathbf{R}(\mathbf{Q}_{n+1}^k) \tag{3.34}$$

is solved and $\Delta\mathbf{Q}_{n+1}^{k+1}$ is used in an additive or multiplicative way depending on the quantities to update $\mathbf{Q}_{n+1}^k$ until a given convergence criteria is met. Here, a very common choice is made to assess whether the computation is converged. It is demanded that

$$||\mathbf{R}(\mathbf{Q}_{n+1}^{k+1})|| < \text{error}_{\mathbf{R}} \text{ and } ||\Delta\mathbf{Q}_{n+1}^{k+1}|| < \text{error}_{\Delta\mathbf{Q}}, \tag{3.35}$$

where the thresholds $\text{error}_{\mathbf{R}}$ and $\text{error}_{\Delta\mathbf{Q}}$ are chosen to fit the requirements and dimension of the computation.

If the initial guess, which is chosen in the present work to be the last converged result, is close enough to the actual solution, the Newton-Raphson method exhibits a favorable quadratic convergence order. Initial estimates of dynamic quantities are also taken from the last converged solution. If the numerical procedure for a conservative (quasi-)static problem does not converge due to an inadequate initial guess that is not close enough to the solution the external loads can be gradually increased. In the present work the load step adaption scheme from Meier et al. (2017) is employed to allow for comparison of required Newton-steps in Sect. 3.6: A tenth of the load is applied at the beginning. In case that the computation diverges, the load increment is halved. Conversely, when four successive load increases have converged, the load increment is doubled.

A large number of alternative solution strategies are available trying to increase efficiency and robustness of the numerical procedure. Some prominent examples are named in the following but for further studies the reader is referred to e.g. Deuflhard (2004), Kelley (2003), Nocedal & Wright (1999), and Ortega & Rheinboldt (1970). In so-called line-search variants of the Newton-Raphson scheme $\Delta\mathbf{Q}_{n+1}^{k+1}$ is scaled in the update procedure (Gill & Murray, 1974; Lee, 1989). In quasi-Newton methods the stiffness matrix is modified in various ways, for example in pseudo-transient continuation schemes (e.g. Ceze & Fidkowski 2015; Gee et al. 2009; Kelley & Keyes 1998). In unstable systems numerical procedures with elaborate step-size control are commonly applied, such as arc-length variants (e.g. Crisfield 1983; Crisfield 1981; Ritto-Corrêa & Camotim 2008).

Usually, solving the linear system of Eq. (3.34) poses the biggest computational effort in a finite-element procedure where the nature of the global stiffness matrix depends on the underlying physical problem. At the same time it determines the crucial properties of the solving process. Direct solvers based on a LU-decomposition might be applied to small systems. For bigger systems iterative solvers such as the conjugate gradient method (Hestenes & Stiefel, 1952) or the generalized minimal residual method (Saad, 2003) are preferable. The performance of these methods are strongly influenced by the ratio of the largest to the smallest eigenmode of $\mathbf{K}$. To reduce the ratio and improve the performance preconditioning schemes, such as incomplete factorization, algebraic multi-grid methods, and Jacobi and Gauss-Seidel methods, might be employed. Alternatively, when these strategies fail, the underlying physical model needs to be altered, e.g. removing shear and axial deformation modes associated with one of the extreme eigenmodes. In Quarteroni et al. (2010) more details on the topic of solving large systems of equations can be found.

Mixed finite-element formulations: Typically, kinematic quantities describing displacement and rotations are the primary unknown fields, as example the beam formulation presented in Sect. 2.5.3 can be given. But in so-called mixed or hybrid finite-element formulations additional fields are introduced, mostly for numerical benefits. The new unknown fields can be employed to increase accuracy and avoid ill-conditioning and locking phenomena (Arunakirinathar & Reddy, 1993; Frischkorn & Reese, 2013; Sansour et al., 1996). In general, mixed variational principles of Hellinger-Reissner type (Hellinger, 1907; Reissner, 1950), where displacements, rotations, and stresses are the variables, and of Hu-Wahizu type (Hu, 1955; Washizu, 1955), where in addition strains are introduced as an unknown field, can be identified. Furthermore, some mixed formulations require to restrict some of the primary fields using Lagrangian multipliers causing the use of additional nodal degrees of freedom, see Sect. 2.5.5.3.

The literature on geometrically exact beams may also be grouped by their choice of primary fields and the usage of Lagrangian multipliers. Referring to the aforementioned types, a 2-field Hellinger-Reissner type formulation was presented by Nukala & White (2004) as well as a 3-field Hu-Wahizu type formulation that was first devised by Iura & Atluri (1988a), Iura & Atluri (1988b) and Pimenta & Yojo (1993). Hodges (1990) and Santos et al. (2010) provided good overviews over mixed-field beam formulations but without implementation in the finite-element framework.

There are different strategies to circumvent the necessity of multipliers. In so-called fully intrinsic approaches, Hegemier & Nair (1977), Hodges (2003), and Sotoudeh & Hodges (2011) excluded any finite rotation or displacement variables but included not only strains but also resultant forces and momenta as independent quantities. Similarly, but with restriction to straight reference configurations, Jelenić & Saje (1995) required only rotational degrees of freedom.

By using strain fields to implicitly define kinematic fields Zupan & Saje (2003, 2006) obtained locking-free formulations but to compute the total rotations from the strains a numerical integration scheme had to be applied. Češarek et al. (2012) avoided this problem by assuming linear strains within each element ultimately reducing the convergence order of the formulation. A comparable approach was chosen in Sect. 2.5.2.

3.2.2 Discretization of Elements of $\mathbb{R}$ *and* $\mathbb{R}^3$

In this section discretizations of the axial strain ε, the center line $\boldsymbol{r}$, and the tangent field $\boldsymbol{t}$ are presented and different interpolation strategies for the center line are compared. Depending on the consistency requirements resulting from the underlying weak form C^0- and C^1-continuous interpolations are required. More precisely, only the Kirchhoff-Love beam based on constraint rotation presented in Sect. 2.5.2 requires C^1-continuity. To this end, Lagrange shape functions L_o of order o and Hermite shape functions are employed in the following. It is repeated that nodal values of the primary fields appear as $(\hat{\bullet})_i$ with $i = 1, ..., n_{\text{node}}$. Naturally, the Lagrange shape functions are of order $(n_{\text{node}} - 1)$. From the discretization of the center line an expression for the element Jacobian can be specified to be

$$J^e(\xi) := \left\|\frac{\mathrm{d}\boldsymbol{r}_{0h}}{\mathrm{d}\xi}\right\| \quad \text{such that} \quad \mathrm{d}s_0(\xi) = J^e(\xi)\mathrm{d}\xi. \tag{3.36}$$

The element Jacobian links the differentials with respect to the local and global coordinates. It varies over the undescretized center line and simply employing a constant value for the Jacobian for all elements might affect the consistency of the finite element formulation. According to Meier et al. (2014), different reasonable definitions for the element length h_e can be formulated, all of which need to minimize the result of the integral

$$\int_0^{L_0} (\boldsymbol{r}_0 - \boldsymbol{r}_{0,h})^2 \mathrm{d}s_0. \tag{3.37}$$

Meier et al. (2014) found the optimal choice to be

$$h_e := \int_{-1}^{+1} ||\frac{\partial \boldsymbol{r}_{0,h}(\xi, h_e)}{\partial \xi}||\mathrm{d}\xi, \tag{3.38}$$

which is also used in the present work. If no analytical representation of the initial center line is available, h_e is determined for every element using an iterative solution scheme once at the beginning of a computation (Armero & Valverde, 2012a). The center line is obtained via numerical integration of the tangent field. But it should also be noted that the differences between reasonable choices quickly diminish when refining the mesh.

Discretization of the axial deformation: ε and the according test function $\delta\varepsilon$ are discretized in a Bubnov-Galerkin manner yielding

$$\varepsilon_h^e(\xi) = \sum_{i=1}^{n_{node}} L_i(\xi)\hat{\varepsilon}_i(\xi) := \mathrm{L}^e\hat{\boldsymbol{\varepsilon}}^e \text{ and } \delta\varepsilon_h^e = \sum_{i=1}^{n_{node}} L_i(\xi)\delta\hat{\varepsilon}_i := \mathrm{L}^e\delta\hat{\boldsymbol{\varepsilon}}^e, \tag{3.39}$$

where L^e, $\hat{\boldsymbol{\varepsilon}}^e$, and $\delta\hat{\boldsymbol{\varepsilon}}^e$ are proper assemblies on element level of the shape functions and nodal values, i.e.

$$\hat{\boldsymbol{\varepsilon}}^e := \left(\hat{\varepsilon}_1, ..., \hat{\varepsilon}_{n_{node}}\right)^T \text{ and } \delta\hat{\boldsymbol{\varepsilon}}^e := \left(\delta\hat{\varepsilon}_1, ..., \delta\hat{\varepsilon}_{n_{node}}\right)^T \text{ with } \hat{\varepsilon}_i, \delta\hat{\varepsilon}_i \in \mathbb{R} \tag{3.40}$$

and

$$\mathrm{L}^e := (L_1(\xi), ..., L_{n_{\text{node}}}(\xi)). \tag{3.41}$$

L_i denotes the standard Lagrange shape functions, where at the node j with the local coordinate ξ_j it holds that $L_i(\xi_j) = \delta_{ij}$. In the present work, $n_{\text{node}} = 3$ is chosen most commonly such that

$$L_1(\xi) = \frac{1}{2}\xi(\xi - 1), \quad L_2(\xi) = 1 - \xi^2, \quad \text{and} \quad L_3(\xi) = \frac{1}{2}\xi(\xi + 1). \tag{3.42}$$

In later sections derivatives with regard to ξ are required and for the sake of conciseness

$$\mathrm{L}^e_\xi := \left(\frac{\mathrm{d}L_1(\xi)}{\mathrm{d}\xi}, \ldots, \frac{\mathrm{d}L_{\text{node}}(\xi)}{\mathrm{d}\xi}\right) \tag{3.43}$$

is defined. The subscript $(\bullet)_\xi$ signifies a partial derivative with respect to ξ.

Direct discretization of the beam center-line using Lagrange shape functions: Similar to the last section, $\boldsymbol{r}$ and the according test function $\delta\boldsymbol{r}$ are discretized using Lagrange shape functions yielding

$$\boldsymbol{r}^e_h = \sum_{i=1}^{n_{node}} L_i(\xi)\hat{\boldsymbol{r}}_i := \mathbf{L}^e\hat{\mathbf{r}}^e \text{ and } \delta\boldsymbol{r}^e_h = \sum_{i=1}^{n_{node}} L_i(\xi)\delta\hat{\boldsymbol{r}}_i := \mathbf{L}^e, \delta\hat{\mathbf{r}}^e, \tag{3.44}$$

where

$$\hat{\mathbf{r}}^e := \left(\hat{\boldsymbol{r}}^T_1, \ldots, \hat{\boldsymbol{r}}^T_{n_{node}}\right)^T \text{ and } \delta\hat{\mathbf{r}}^e := \left(\delta\hat{\boldsymbol{r}}^T_1, \ldots, \delta\hat{\boldsymbol{r}}^T_{n_{node}}\right)^T \text{ with } \hat{\boldsymbol{r}}_i, \delta\hat{\boldsymbol{r}}_i \in \mathbb{R}^3 \tag{3.45}$$

and

$$\mathbf{L}^e := (L_1(\xi)\boldsymbol{I}_{3\times3}, \ldots, L_{n_{\text{node}}}(\xi)\boldsymbol{I}_{3\times3}). \tag{3.46}$$

Herein, $\boldsymbol{I}_{3\times3}$ denotes the 3×3 identity tensor. In later sections derivatives with regard to ξ are required and for concise notation

$$\mathbf{L}^e_\xi := \left(\frac{\mathrm{d}L_1(\xi)}{\mathrm{d}\xi}\boldsymbol{I}_{3\times3}, \ldots, \frac{\mathrm{d}L_{\text{node}}(\xi)}{\mathrm{d}\xi}\boldsymbol{I}_{3\times3}\right). \tag{3.47}$$

is defined.

Direct discretization of the beam center-line using Hermite shape functions: The Kirchhoff-Love beam based on constraint rotation presented in Sect. 2.5.3 requires C^1-continuity. Therefore, the Hermite shape functions

$$\begin{aligned} H^r_1(\xi) &= \frac{1}{4}(2+\xi)(1-\xi)^2, \; H^r_2(\xi) = \frac{1}{4}(2-\xi)(1+\xi)^2, \\ H^t_1(\xi) &= \frac{1}{4}(1+\xi)(1-\xi)^2, \; H^t_2(\xi) = -\frac{1}{4}(1-\xi)(1+\xi)^2 \end{aligned} \tag{3.48}$$

are employed. Naturally, they exactly represent the interpolated quantities at the nodes

$$H_i^r(\xi_j) = \delta_{ij}, \quad H_{i,\xi}^r(\xi_j) = 0, \quad H_i^t(\xi_j) = \delta_{ij}, \quad H_{i,\xi}^t(\xi_j) = 0 \text{ for } i, j = 1, 2 \quad (3.49)$$

and they meet the completeness conditions

$$\sum_{i=1}^{2} H_i^r(\xi) = 1, \quad \sum_{i=1}^{2}(\xi_i H_i^r(\xi) + H_i^t(\xi)) = \xi,$$
$$\sum_{i=1}^{2}(\xi_i^2 H_i^r(\xi) + 2\xi_i H_i^t(\xi)) = \xi^2, \quad \sum_{i=1}^{2}(\xi_i^3 H_i^r(\xi) + 3\xi_i^2 H_i^t(\xi)) = \xi^3. \quad (3.50)$$

required to exactly interpolate polynomials of third order. The subscript $(\bullet)_\xi$ signifies a partial derivative with respect to ξ. Assembling the Hermite shape functions according to

$$\mathbf{H}^e = \left(H_1^r(\xi)\boldsymbol{I}_{3\times3}, \frac{c}{2}H_1^t(\xi)\boldsymbol{I}_{3\times3}, H_2^r(\xi)\boldsymbol{I}_{3\times3}, \frac{c}{2}H_2^t(\xi)\boldsymbol{I}_{3\times3}\right), \quad (3.51)$$

the center line can be discretized using two nodes at the element boundary to obtain

$$\boldsymbol{r}_h(\xi) = \sum_{i=1}^{2} H_i^r(\xi)\hat{\boldsymbol{r}}_i + \frac{c}{2}\sum_{i=1}^{2} H_i^t(\xi)\hat{\boldsymbol{h}}_i =: \mathbf{H}^e\hat{\mathbf{h}}^e \quad (3.52)$$

and in a Bubnov-Galerkin approach the test function is discretized using the same Hermite functions

$$\delta\boldsymbol{r}_h(\xi) = \sum_{i=1}^{2} H_i^r(\xi)\delta\hat{\boldsymbol{r}}_i + \frac{c}{2}\sum_{i=1}^{2} H_i^t(\xi)\delta\hat{\boldsymbol{t}}_i =: \mathbf{H}^e\delta\hat{\mathbf{h}}^e. \quad (3.53)$$

Herein, $\hat{\boldsymbol{r}}_i$ and $\hat{\boldsymbol{t}}_i$ are the nodal position and tangent vectors, respectively, and $\delta\hat{\boldsymbol{r}}_i$ and $\delta\hat{\boldsymbol{t}}_i$ denote the respective weighting functions. $c > 0$ is a constant related to the element length h_e with the physical unit of length, which should be chosen to optimally represent the discretized center line, see the beginning of this Section. For a two-node element, as used in the present work, these quantities are assembled to

$$\hat{\mathbf{h}}^e = \left(\hat{\boldsymbol{r}}_1^T, \hat{\boldsymbol{t}}_1^T, \hat{\boldsymbol{r}}_2^T, \hat{\boldsymbol{t}}_2^T\right)^T \text{ and } \delta\hat{\mathbf{h}}^e = \left(\delta\hat{\boldsymbol{r}}_1^T, \delta\hat{\boldsymbol{t}}_1^T, \delta\hat{\boldsymbol{r}}_2^T, \delta\hat{\boldsymbol{t}}_2^T\right)^T \text{ with } \hat{\boldsymbol{r}}_i, \delta\hat{\boldsymbol{r}}_i, \hat{\boldsymbol{t}}_i, \delta\hat{\boldsymbol{t}}_i \in \mathbb{R}^3. \quad (3.54)$$

Furthermore, for a concise notation of derivatives

$$\mathbf{H}^e_\xi = \left(\frac{\mathrm{d}H^r_1(\xi)}{\mathrm{d}\xi} \boldsymbol{I}_{3\times 3}, \frac{c}{2}\frac{\mathrm{d}H^t_1(\xi)}{\mathrm{d}\xi} \boldsymbol{I}_{3\times 3}, \frac{\mathrm{d}H^r_2(\xi)}{\mathrm{d}\xi} \boldsymbol{I}_{3\times 3}, \frac{c}{2}\frac{\mathrm{d}H^t_2(\xi)}{\mathrm{d}\xi} \boldsymbol{I}_{3\times 3} \right) \tag{3.55}$$

is defined.

Indirect discretization of the beam center-line using its tangent field: The implicit Kirchhoff-Love beam based on constraint translation presented in Sect. 2.5.2 requires only C^0-continuity. Therefore, the tangent field is represented using Lagrangian shape functions yielding

$$\boldsymbol{t}^e_h = \sum_{i=1}^{n_{node}} L_i(\xi)\hat{\boldsymbol{t}}_i := \mathbf{L}^e\hat{\mathbf{t}}^e \text{ and } \delta\boldsymbol{t}^e_h = \sum_{i=1}^{n_{node}} L_i(\xi)\delta\hat{\boldsymbol{t}}_i := \mathbf{L}^e\delta\hat{\mathbf{t}}^e, \tag{3.56}$$

where

$$\hat{\mathbf{t}}^e = \left(\hat{\boldsymbol{t}}_i^T, \dots, \hat{\boldsymbol{t}}_{n_{node}}^T\right)^T \text{ and } \delta\hat{\mathbf{t}}^e = \left(\delta\hat{\boldsymbol{t}}_i^T, \dots, \delta\hat{\boldsymbol{t}}_{n_{node}}^T\right)^T \text{ with } \hat{\boldsymbol{t}}_i, \delta\hat{\boldsymbol{t}}_i \in \mathbb{R}^3. \tag{3.57}$$

The actual center line position has to be computed via an integral over all elements starting from a known position according to Eq. (2.120) such that

$$\boldsymbol{r}_h(s_0) = \boldsymbol{r}^*(s_0^*) + \int_{s_0^*}^{s_0} \boldsymbol{t}_h \mathrm{d}s_0. \tag{3.58}$$

Discretization of Velocity, Acceleration, and Increments: The quantities $\dot{\varepsilon}$, $\ddot{\varepsilon}$, $\dot{\boldsymbol{r}}$, $\ddot{\boldsymbol{r}}$, $\dot{\boldsymbol{t}}$, $\ddot{\boldsymbol{t}}$, $\Delta\varepsilon$, $\Delta\varepsilon'$, $\Delta\boldsymbol{r}$, $\Delta\boldsymbol{r}'$, $\Delta\boldsymbol{t}$, and $\Delta\boldsymbol{t}'$ need to be discretized as well, where the discretized quantities are denoted by the usual index $(\bullet)_h$. The tedious notations of the last sections are not repeated. More detailed definitions are given where required.

In the scope of the present work the velocity and acceleration vectors require only C^0-continuity and are interpolated based on their nodal values using standard Lagrangian shape functions of order $(n_{\text{node}} - 1)$ for n_{node}-noded elements as done in the previous sections. Their nodal values need to be updated according to Sect. 3.1.2.

The increments and their derivatives can be realized in different ways using either Lagrangian or Hermite shape functions. For increments of translational quantities the interpolation strategy of the primary field is used.

Accuracy and efficiency of different center-line discretizations: In this paragraph, the different center line discretization strategies defined in Eqs. (3.44), (3.52), and (3.56) are compared with regard to accuracy and efficiency. In detail, a study of the error in dependence of the number of degrees of freedom n_{dof} for a space curve is performed. A quarter circle is studied on the basis of the L^2-error

$$||e||^2_r = \sqrt{\frac{1}{L_0}\int_0^{L_0} ||\boldsymbol{r} - \boldsymbol{r}_h||^2 \mathrm{d}s_0}, \tag{3.59}$$

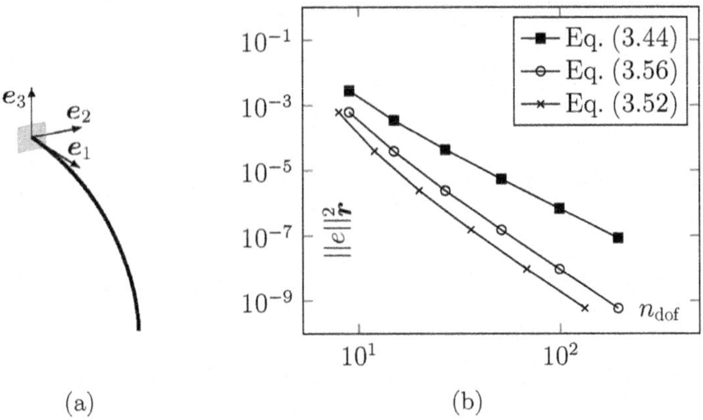

Fig. 3.1 Study of the accuracy and efficiency of different center line discretizations on the basis of a quarter circle (**a**). The L^2-error $\|e\|_r^2$ is shown over the number of degrees of freedom n_{dof} (**b**)

where L_0 is the total (undeformed) length of the center line. The result is shown in Fig. 3.1. The highest efficiency is found using Hermite shape functions as defined in Eq. (3.52). The most interesting result is that the implicit discretization from Eq. (3.56) vastly outperforms the explicit discretization from Eq. (3.44). This is due to the smoothing effect of the integral involved. In Sect. 3.6 beam elements are compared, which are based on these different center line discretizations. But despite the result obtained here, the Kirchhoff-Love beams using the discretization detailed in Eq. (3.56) shows the best performance. However, not all possible combinations of weak form and discretization methods for the center line and the rotations were tested and an even better formulation might still emerge in future research.

3.2.3 Discretization of Elements of $SO(3)$

In the following the discretization strategies based on the rotation parameterization as set out in Sects. 2.3.3.1 and 2.3.3.2 are detailed. These approaches were chosen based on the arguments presented in Sect. 2.3.3.3. While doing so, the difference between parameterization and interpolation should be stressed. The parameterization determines the quantities stored at the nodes presetting the corresponding update procedure. The interpolation strategy specifies how a rotation tensor is obtained apart from the nodal values. This section is concerned with the latter.

Discretization based on local rotation vector: The first approach to the interpolation of ε and $\boldsymbol{\Lambda}$ is based on Sect. 2.3.3 and the approved strategy developed by Jelenić & Crisfield (1999). Incorporating their important results regarding objectivity and path-independence the discretized local rotation vector

$$\boldsymbol{\psi}_h^l(\xi) = \sum_{i=1}^{n_{node}} L_i(\xi)\hat{\boldsymbol{\psi}}_i^l := \mathbf{L}^e\hat{\boldsymbol{\psi}}^l \tag{3.60}$$

is formulated with

$$\hat{\boldsymbol{\psi}}^l := \left(\left(\hat{\boldsymbol{\psi}}_1^l\right)^T, ..., \left(\hat{\boldsymbol{\psi}}_{n_{node}}^l\right)^T\right)^T \quad \text{and} \quad \hat{\boldsymbol{\psi}}_i^l \in \mathbb{R}^3 \tag{3.61}$$

such that the transformation tensor

$$\boldsymbol{\Lambda}_h(\xi) = \boldsymbol{\Lambda}_r \ \exp(\boldsymbol{S}(\boldsymbol{\psi}_h^l)) \equiv \boldsymbol{\Lambda}_r \left(\boldsymbol{I} + \frac{\sin\Psi}{\Psi}\boldsymbol{S}(\boldsymbol{\psi}_h^l) + \frac{1-\cos\Psi}{\Psi^2}\boldsymbol{S}(\boldsymbol{\psi}_h^l)^2\right) \tag{3.62}$$

can be obtained in its discretized form with $\Psi = ||\boldsymbol{\psi}_h^l||$. As firstly proposed by Jelenić & Crisfield (1999) the reference rotation tensor $\boldsymbol{\Lambda}_r$ is chosen as the 'midway' rotation. It is obtained by first determining the rotation vector, which takes $\boldsymbol{\Lambda}_1$ to $\boldsymbol{\Lambda}_{n_{node}}$ and secondly applying half of that vector to $\boldsymbol{\Lambda}_1$.

For the discretization scheme to be applied within a finite element framework, increments of the relative rotation vector $\Delta\boldsymbol{\psi}^l$ need to be related to $\Delta\boldsymbol{\theta}$. To this end, Jelenić & Crisfield (1999) were able to derive so-called generalized interpolation functions $\tilde{\boldsymbol{L}}_i(\xi)$. The corresponding lengthy derivation of the respective derivatives can be found therein. Using these functions does not only preserve objectivity and path-independence of the beam formulation but also $\Delta\boldsymbol{\theta}$ can be conveniently approximated in terms of its nodal values $\Delta\boldsymbol{\theta}_i$ in a familiar manner as

$$\Delta\boldsymbol{\theta}_h^e(\xi) = \sum_{i=1}^{n_{node}} \tilde{\boldsymbol{L}}_i(\xi)\Delta\hat{\boldsymbol{\theta}}_i := \tilde{\mathbf{L}}^e\Delta\hat{\boldsymbol{\theta}}^e, \tag{3.63}$$

where

$$\Delta\hat{\boldsymbol{\theta}}^e := \left(\Delta\hat{\boldsymbol{\theta}}_1^T, ..., \Delta\hat{\boldsymbol{\theta}}_{n_{node}}^T\right)^T \quad \text{and} \quad \tilde{\mathbf{L}}^e := \left(\tilde{\boldsymbol{L}}_1, ..., \tilde{\boldsymbol{L}}_{n_{node}}\right) \tag{3.64}$$

with $\Delta\hat{\boldsymbol{\theta}}_i \in \mathbb{R}^3$ and $\tilde{\boldsymbol{L}}_i \in \mathbb{R}^3 \times \mathbb{R}^3$.

Furthermore, a Petrov-Galerkin approach is chosen in order to avoid additional linearization resulting in

$$\delta\boldsymbol{\theta}_h(\xi) = \sum_{i=1}^{3} L_i(\xi)\delta\hat{\boldsymbol{\theta}}_i := \mathbf{L}^e\delta\hat{\boldsymbol{\theta}}^e \ \text{ with } \ \delta\hat{\boldsymbol{\theta}}^e = \left(\delta\hat{\boldsymbol{\theta}}_i^T, ..., \delta\hat{\boldsymbol{\theta}}_{n_{node}}^T\right)^T. \tag{3.65}$$

It is worthwhile noting that the discretized material curvature vector

$$\boldsymbol{K}_h = \left(\boldsymbol{T}^{-1}\left(\boldsymbol{\psi}_h^l(\xi)\right)\right)^T \frac{\mathrm{d}\boldsymbol{\psi}_h^l(\xi)}{\mathrm{d}s_0} \tag{3.66}$$

can be expressed using the operator defined in Eq. (2.43). Since the presented discretization is C^0-continuous, a state of constant curvature can be exactly represented. Thus, as frequently mentioned in the literature, see e.g. Sander (2010), the presented method is a geodesic interpolation scheme. It connects two elements of $SO(3)$ via a straight line defined in its tangent space.

Discretization based on smallest-rotation split: Based on the results of Sect. 2.3.3.2 $\mathbf{\Lambda}_h$ is interpolated within each element yielding expressions closely related to those available in Meier et al. (2014, 2017). The tangent vector field, a local twist parameter φ^l, and their corresponding test functions are discretized in a Bubnov-Galerkin manner, see Sect. 3.5.4 for a justification, to obtain

$$\boldsymbol{t}_h^e = \sum_{i=1}^{n_{node}} L_i(\xi)\hat{\boldsymbol{t}}_i := \mathbf{L}^e\hat{\mathbf{t}}^e \quad \text{and} \quad \varphi_h^l = \sum_{i=1}^{n_{node}} L_i(\xi)\hat{\varphi}_i^l := \mathrm{L}^e\hat{\varphi}^l, \tag{3.67}$$

and

$$\delta\boldsymbol{t}_h^e = \sum_{i=1}^{n_{node}} L_i(\xi)\delta\hat{\boldsymbol{t}}_i := \mathbf{L}^e\delta\hat{\mathbf{t}}^e \quad \text{and} \quad \delta\varphi_h^l = \sum_{i=1}^{n_{node}} L_i(\xi)\delta\hat{\varphi}_i^l := \mathrm{L}^e\delta\hat{\varphi}^l, \tag{3.68}$$

respectively. Therein, the assembled quantities

$$\varphi_h^l := \left(\hat{\varphi}_1^l, ..., \hat{\varphi}_{n_{node}}^l\right)^T \quad \text{and} \quad \delta\varphi_h^l := \left(\delta\hat{\varphi}_1^l, ..., \delta\hat{\varphi}_{n_{node}}^l\right)^T \quad \text{with} \quad \hat{\varphi}_i^l, \delta\hat{\varphi}_i^l \in \mathbb{R} \tag{3.69}$$

are used. The nodal values of φ_i^l are determined by solving

$$\exp(\boldsymbol{S}(\varphi_i^l\boldsymbol{g}_{1_i})) = \mathbf{\Lambda}_i\mathbf{\Lambda}_r^T\mathbf{\Lambda}_{sr_i}^T \quad \text{with} \quad \boldsymbol{g}_{1_i} = \frac{\boldsymbol{t}_i}{||\boldsymbol{t}_i||} \tag{3.70}$$

for φ_i^l. Further choosing the reference rotation tensor to be the one at node J such that $\mathbf{\Lambda}_r = \mathbf{\Lambda}_J$ and defining $\boldsymbol{R}_h = \exp(\boldsymbol{S}(\varphi_h^l\boldsymbol{g}_{1_h}))$, the discretized rotation tensor can be given as

$$\mathbf{\Lambda}_h(\xi) = \boldsymbol{R}_h\mathbf{\Lambda}_{sr_h}\mathbf{\Lambda}_J, \tag{3.71}$$

respectively. In contrast to Sect. 3.2.3, a Bubnov-Galerkin approach is employed. Although it leads to additional linearizations, it is necessary to be able to represent a constant rotation vector over the element (Meier et al., 2017).

It is important to note the difference between φ, whose increments are used to update $\mathbf{\Lambda}_i$, see Eq. (2.88), and φ^l, which is a quantity computed in each element to interpolate $\mathbf{\Lambda}_h$.

Consequently, building on Eqs. (2.73) and (2.74) the variation of a single nodal rotation tensor,

$$\delta \boldsymbol{\Lambda}_i = \boldsymbol{S}(\delta\boldsymbol{\theta}_i + \delta\varphi_i \boldsymbol{g}_{1_i})\boldsymbol{\Lambda}_i \text{ with } \delta\boldsymbol{\theta}_i = \boldsymbol{T}_i\delta\boldsymbol{t}_i \text{ and } \boldsymbol{T}_i = \frac{\boldsymbol{S}(\boldsymbol{g}_{1_i})}{||\boldsymbol{t}_i||} \quad (3.72)$$

can be obtained. Using this expression for $\delta\boldsymbol{\Lambda}_i$ in combination with Eq. (2.74), the variation of $\boldsymbol{\Lambda}_h$ reads

$$\begin{aligned}\delta\boldsymbol{\Lambda}_h =& \boldsymbol{S}(\delta\boldsymbol{\theta}_{\delta t})\boldsymbol{\Lambda}_h + \boldsymbol{S}(\delta\boldsymbol{\theta}_{\delta\varphi^l})\boldsymbol{\Lambda}_h \\ &+ \delta_{iJ}\left[\boldsymbol{R}_h\boldsymbol{S}(\delta\boldsymbol{\theta}_{\delta t_i})\boldsymbol{\Lambda}_{sr_h}\boldsymbol{\Lambda}_i + \boldsymbol{R}_h\boldsymbol{\Lambda}_{sr_h}\boldsymbol{S}(\delta\boldsymbol{\theta}_i)\boldsymbol{\Lambda}_i\right].\end{aligned} \quad (3.73)$$

Therein, the relations

$$\boldsymbol{\Lambda}_r = \boldsymbol{\Lambda}_J \text{ and } \delta\boldsymbol{\theta}_{\delta\varphi^l} = \delta\varphi_h^l \boldsymbol{g}_{1_h} \text{ with } \boldsymbol{g}_{1_h}(\xi) = \frac{\boldsymbol{t}_h(\xi)}{||\boldsymbol{t}_h(\xi)||} \quad (3.74)$$

are employed. Also, $\delta\boldsymbol{\theta}_{\delta t_i}$ is defined by replacing $\boldsymbol{t}_0$ with $\boldsymbol{t}_i$ in Eq. (2.75).
In a subsequent step, Eq. (3.73) is reformulated by multiplying $\boldsymbol{\Lambda}_h^{-1}$ from the right-hand side and using the general relation $\boldsymbol{S}(\boldsymbol{A}\boldsymbol{a}) = \boldsymbol{A}\boldsymbol{S}(\boldsymbol{a})\boldsymbol{A}^T$ for $\boldsymbol{a} \in \mathbb{R}$ and $\boldsymbol{A} \in \mathbb{R}^3 \times \mathbb{R}^3$ to obtain

$$\delta\boldsymbol{\theta}_h = \delta\boldsymbol{\theta}_{\delta t} + \delta\boldsymbol{\theta}_{\delta\varphi^l} + \delta_{iJ}\left[\boldsymbol{R}_h\delta\boldsymbol{\theta}_{\delta t_i} + \boldsymbol{R}_h\boldsymbol{\Lambda}_{sr_h}\delta\boldsymbol{\theta}_i\right]. \quad (3.75)$$

To be useful, $\delta\varphi_i^l$ needs to be expressed in terms of $\delta\boldsymbol{t}_i$ and $\delta\varphi_i$ in a final step. For this purpose, it is determined what amount of twist about $\boldsymbol{g}_i$ that arises from a variation of $\boldsymbol{\Lambda}_i$ according to Eq. (3.72) is included by a variation solely due to $\delta\boldsymbol{t}$. The difference is then represented by the nodal local twist variation

$$\delta\varphi_i^l = \boldsymbol{g}_{1_i}^T(\delta\boldsymbol{\theta}_i + \delta\varphi_i\boldsymbol{g}_{1_i}) - \boldsymbol{g}_{1_i}^T\left(\delta\boldsymbol{\theta}_{\delta t} + \delta_{iJ}\left[\boldsymbol{R}\delta\boldsymbol{\theta}_{\delta t_i} + \boldsymbol{R}\boldsymbol{\Lambda}_{sr}\delta\boldsymbol{\theta}_i\right]\right), \quad (3.76)$$

which is reformulated to be used in Eq. (3.74) as

$$\delta\varphi_i^l\boldsymbol{g}_{1_h} = \frac{\boldsymbol{g}_{1_h} \otimes (\boldsymbol{g}_{1_J} \times \boldsymbol{g}_{1_i})}{||\boldsymbol{t}_i||(1 + \boldsymbol{g}_{1_i}^T\boldsymbol{g}_{1_J})}\delta\boldsymbol{t}_i - \frac{\boldsymbol{g}_{1_h} \otimes (\boldsymbol{g}_{1_i} \times \boldsymbol{g}_{1_J})}{||\boldsymbol{t}_J||(1 + \boldsymbol{g}_{1_i}^T\boldsymbol{g}_{1_J})}\delta\boldsymbol{t}_J + \delta\varphi_i\boldsymbol{g}_{1_h}. \quad (3.77)$$

The quantities $\dot{\boldsymbol{W}}$, $\ddot{\boldsymbol{W}}$, $\dot{\boldsymbol{A}}$, and $\ddot{\boldsymbol{A}}$ require only C^0-continuity and are interpolated using standard Lagrangian shape functions of order $n_{\text{node}} - 1$ for n_{node}-noded elements. The discretized versions are signified as $\dot{\boldsymbol{W}}_h$, $\ddot{\boldsymbol{W}}_h$, $\dot{\boldsymbol{A}}_h$, and $\ddot{\boldsymbol{A}}_h$. The nodal values need to be updated according to Sect. 3.1.3.

The rotation increment $\Delta\boldsymbol{\theta}$ requires special generalized shape functions detailed in Eq. (3.63) and the twist increment $\Delta\varphi$ is discretized using Lagrangian shape functions. See Sect. 2.3.3.2 for the corresponding update procedure of the nodal values.

The different rotation discretization strategies as proposed in Eqs. (3.62) and (3.71) are compared exhaustively with regard to accuracy and efficiency in Sect. 3.6.

Discretization of Lagrangian multipliers A weak enforcement of the Kirchhoff-Love elements, as proposed in Sect. 3.4.3, involves the discretization of Lagrangian multipliers reading

$$\boldsymbol{\lambda}_h^e := \mathbf{L}^e \hat{\boldsymbol{\lambda}}^e, \tag{3.78}$$

where $\boldsymbol{\lambda}_h^e = [\lambda_{2,h}^e, \lambda_{3,h}^e]^T$ and $\hat{\boldsymbol{\lambda}}^e = [\hat{\lambda}_2, \hat{\lambda}_3]^T$ are the vectors containing the discretized multipliers and nodal values, respectively. The discretized variations and increments of the multipliers result analogously.

3.3 Simo-Reissner Beam Element

For completeness and later references as well as numerical comparisons, the weak form of the Simo-Reissner beam as derived in Sect. 2.4.3 is presented in its discretized form closely related to the work by Jelenić & Crisfield (1999) giving rise to the element residual vector $\mathbf{r}_{\mathrm{JC}}^{\mathrm{e}}$ and the element stiffness matrix $\mathbf{k}_{\mathrm{JC}}^{\mathrm{e}}$. The beam domain is subdivided into a total number of n_{e} element domains B_h^{e} with n_{node} nodes each. The highest occurring derivative of an unknown field is of first order calling for C^0-continuity at the element borders. Accordingly, the trial subspace $(\boldsymbol{r}_h, \boldsymbol{\Lambda}_h) \in \mathcal{T}_h$ and the weighting subspace $(\delta \boldsymbol{r}_h, \delta \boldsymbol{\theta}_h) \in \mathcal{W}_h$ are constructed. The center-line is discretized using Lagrangian shape functions according to Sect. 3.2.2 and the rotation tensor is discretized using the generalized shape functions in a Petrov-Galerkin manner as defined in Sect. 3.2.3 such that the trail subspace could also be constructed as $(\boldsymbol{r}_h, \boldsymbol{\psi}_h^l) \in \mathcal{T}_h$. All employed shape functions are of order $n_{\mathrm{node}} - 1$. Furthermore, the temporal discretization techniques from Sects. 3.1.2 and 3.1.3 are used, which is the only difference of the current presentation to its original derivation (Jelenić & Crisfield, 1999). The vectors of nodal values of the trail and test functions and the increments $\hat{\mathbf{q}}_{\mathrm{JC}}^e$, $\delta\hat{\mathbf{q}}_{\mathrm{JC}}^e$, and $\Delta\hat{\mathbf{q}}_{\mathrm{JC}}^e$, respectively, are analogously indexed. The element will be referred to as JC-element.

The original Simo-Reissner beam formulation (Simo, 1985) is also used in comparative examples and is denoted as SR-element but for a detailed description the reader is referred to the corresponding literature.

In the following the superscript $(\bullet)^e$ denoting element quantities and any subscripts related to the time or Newton-iteration step are dropped under the integral for the sake of legibility. The latter implies that all quantities are evaluated at the current Newton-iteration at the current time step t_{n+1} unless indicated by $(\bullet)_n$ referring to the last converged time step.

Element residual vector: The element residual vectors are split into different contributions according to Eq. (3.33) yielding

$$\left(\delta\hat{\mathbf{q}}^e_{\mathrm{JC}}\right)^T \mathbf{r}^e_{\mathrm{int,JC}} = \int\limits_{\Omega^{\mathrm{el}}_h} \left[\delta\boldsymbol{r}'^T_h \boldsymbol{\Lambda}_h \boldsymbol{N}_h - \delta\boldsymbol{\theta}^T_h \boldsymbol{S}(\boldsymbol{r}'_h)\boldsymbol{\Lambda}_h \boldsymbol{N}_h + \delta\boldsymbol{\theta}'^T_h \boldsymbol{\Lambda}_h \boldsymbol{M}_h\right] \mathrm{d}s_0, \tag{3.79}$$

$$\left(\delta\hat{\mathbf{q}}^e_{\mathrm{JC}}\right)^T \mathbf{r}^e_{\mathrm{kin,JC}} = \int\limits_{\Omega^{\mathrm{el}}_h} \left[\delta\boldsymbol{r}^T_h \rho A \ddot{\boldsymbol{r}}_h + \delta\boldsymbol{\theta}^T_h \boldsymbol{\Lambda}_h[\boldsymbol{S}(\boldsymbol{W}_h)\boldsymbol{C}_\rho \boldsymbol{W}_h + \boldsymbol{C}_\rho \boldsymbol{A}_h]\right] \mathrm{d}s_0, \tag{3.80}$$

$$\left(\delta\hat{\mathbf{q}}^e_{\mathrm{JC}}\right)^T \mathbf{r}^e_{\mathrm{ext,JC}} = \int\limits_{\Omega^{\mathrm{el}}_h} \left[\delta\boldsymbol{r}^T_h \boldsymbol{f} + \delta\boldsymbol{\theta}^T_h \boldsymbol{l}\right] \mathrm{d}s_0 + [\delta\boldsymbol{r}^T_h \bar{\boldsymbol{n}} + \delta\boldsymbol{\theta}^T_h \bar{\boldsymbol{m}}]_{\Gamma_\sigma}. \tag{3.81}$$

To avoid the introduction of new symbols at this point, it is not aimed at providing the most concise formulation. With linear constitutive relations it follows that $\boldsymbol{N}_h = \boldsymbol{C}_N(\boldsymbol{\Lambda}_h \boldsymbol{r}'_h - \boldsymbol{E}_1)$ and $\boldsymbol{M}_h = \boldsymbol{C}_M(\boldsymbol{K}_h - \boldsymbol{K}_{0,h})$. The explicit expression of $\boldsymbol{\Lambda}_h$ and $\boldsymbol{K}_h$ depend on the rotation interpolation scheme, which is chosen here to be based on the local rotation vector. Thus, Eqs. (3.62) and (3.66) are used.

Element stiffness matrix: For the iterative solution procedure the first order derivatives at time t_{n+1} are required while any higher-order derivatives are neglected. The element stiffness matrix contribution of the internal elastic part for the Simo-Reissner beam devised by Jelenić & Crisfield (1999) then reads in compact matrix notation

$$\left(\delta\hat{\mathbf{q}}^e_{\mathrm{JC}}\right)^T \mathbf{k}^e_{\mathrm{int,JC}} \Delta\hat{\mathbf{q}}^e_{\mathrm{JC}} = \tag{3.82}$$

$$\int\limits_{\Omega^{\mathrm{el}}_h} \begin{bmatrix} \delta\boldsymbol{r}'_h \\ \delta\boldsymbol{\theta}_h \\ \delta\boldsymbol{\theta}'_h \end{bmatrix}^T \begin{bmatrix} c_N & c_N\,\boldsymbol{S}(\boldsymbol{r}') - \boldsymbol{S}^*(\boldsymbol{N}_h) & \mathbf{0} \\ \boldsymbol{S}^*(\boldsymbol{N}_h) - \boldsymbol{S}(\boldsymbol{r}'_h)\,c_N & \boldsymbol{S}(\boldsymbol{r}')\left(\boldsymbol{S}^*(\boldsymbol{N}_h) - c_N\,\boldsymbol{S}(\boldsymbol{r}')\right) & \mathbf{0} \\ \mathbf{0} & \boldsymbol{\Lambda}_h\,\boldsymbol{S}(\boldsymbol{M}_h)\,\boldsymbol{\Lambda}^T_h & c_M \end{bmatrix} \begin{bmatrix} \Delta\boldsymbol{r}'_h \\ \Delta\boldsymbol{\theta}_h \\ \Delta\boldsymbol{\theta}'_h \end{bmatrix} \mathrm{d}s_0,$$

where $\boldsymbol{S}^*(\boldsymbol{N}_h) = \boldsymbol{\Lambda}_h \boldsymbol{S}(\boldsymbol{N}_h)\boldsymbol{\Lambda}^T_h$.

Before continuing it is stressed that increments of Newton-Raphson steps are denoted by $\Delta\boldsymbol{r}$ and $\Delta\boldsymbol{\theta}$, whereas increments between the current (possibly unconverged) step at t_{n+1} and the last converged time step are $\Delta\tilde{\boldsymbol{r}}$ and $\Delta\tilde{\boldsymbol{\theta}}$ (or its material counterpart $\Delta\tilde{\boldsymbol{\Theta}}$), see Sects. 3.1.2 and 3.1.3, yielding

$$\left(\delta\hat{\mathbf{q}}^e_{\mathrm{JC}}\right)^T \mathbf{k}^e_{\mathrm{kin,JC}} \Delta\hat{\mathbf{q}}^e_{\mathrm{JC}} = \tag{3.83}$$

$$\int\limits_{\Omega^{\mathrm{el}}_h} \begin{bmatrix} \delta\boldsymbol{r}_h \\ \delta\boldsymbol{\theta}_h \end{bmatrix}^T \begin{bmatrix} \rho A \dfrac{\partial \ddot{\boldsymbol{r}}_h}{\partial \boldsymbol{r}_h} & \mathbf{0} \\ \mathbf{0} & -\boldsymbol{S}(\boldsymbol{\pi}_h) + \boldsymbol{\Lambda}_h \boldsymbol{C}_\rho \dfrac{\partial \boldsymbol{A}_h}{\partial \boldsymbol{\theta}_h} + \boldsymbol{\Lambda}_h\left(\boldsymbol{S}(\boldsymbol{W}_h) - \boldsymbol{S}(\boldsymbol{C}_\rho \boldsymbol{W}_h)\right)\dfrac{\partial \boldsymbol{W}_h}{\partial \boldsymbol{\theta}_h} \end{bmatrix} \begin{bmatrix} \Delta\boldsymbol{r}_h \\ \Delta\boldsymbol{\theta}_h \end{bmatrix} \mathrm{d}s_0,$$

where $\boldsymbol{\pi}_h = \boldsymbol{\Lambda}_h\left[\boldsymbol{S}(\boldsymbol{W}_h)\boldsymbol{C}_\rho \boldsymbol{W}_h + \boldsymbol{C}_\rho \boldsymbol{A}_h\right]$. Therein, the partial differentials of the velocity and accelerations depend on the time discretization strategy implemented. With the finite difference method from Sects. 3.1.2 and 3.1.3 it follows

$$\frac{\partial \ddot{\boldsymbol{r}}_h}{\partial \boldsymbol{r}_h} = \frac{1-\alpha_m}{\beta \Delta t^2 (1-\alpha_f)} \frac{\partial \tilde{\boldsymbol{r}}_h}{\partial \boldsymbol{r}_h} = \frac{1-\alpha_m}{\beta \Delta t^2 (1-\alpha_f)} \boldsymbol{I}, \tag{3.84}$$

$$\frac{\partial \boldsymbol{W}_h}{\partial \boldsymbol{\theta}_h} = \frac{\gamma}{\beta \Delta t} \frac{\partial \tilde{\boldsymbol{\Theta}}_h}{\partial \boldsymbol{\theta}_h} = \frac{\gamma}{\beta \Delta t} \boldsymbol{\Lambda}_{h,n}^T \boldsymbol{T}(\tilde{\boldsymbol{\theta}}_h), \tag{3.85}$$

$$\frac{\partial \boldsymbol{A}_h}{\partial \boldsymbol{\theta}_h} = \frac{1-\alpha_m}{\beta \Delta t^2 (1-\alpha_f)} \frac{\partial \tilde{\boldsymbol{\Theta}}_h}{\partial \boldsymbol{\theta}_h} = \frac{1-\alpha_m}{\beta \Delta t^2 (1-\alpha_f)} \boldsymbol{\Lambda}_{h,n}^T \boldsymbol{T}(\tilde{\boldsymbol{\theta}}_h). \tag{3.86}$$

It can be seen that for a dynamic case distinctively more computational effort is required. Both fields $\boldsymbol{\Lambda}_{h,n+1}$ and $\boldsymbol{\Lambda}_{h,n}$ need to be parameterized using local rotation vectors. In addition $\tilde{\boldsymbol{\theta}}_h$ is obtained using Lagrangian shape functions upon determination of the nodal values according to Eq. (2.39).

If external loads are deformation dependent the derivation of the external residual vector has to be computed as well.

3.4 Kirchhoff-Love Beam Elements

In the following, three of the weak formulations of Kirchhoff-Love beams presented in Sect. 2.5 are used to devise a total of four different finite-element formulations. In particular, two different formulations are presented based on the weak form enforcing constraint translation from Sect. 2.5.2, a new inextensible formulation based on Sect. 2.5.5.1 is introduced, and a formulation weakly enforcing the Kirchhoff constraints based on Sect. 2.5.4. The finite-element formulations of the Kirchhoff-Love beams based on constraint rotation as well as the torsion-free are not repeated here as they are available in the literature (Meier et al., 2017). As done in the last section, the superscript $(\bullet)^e$ denoting element quantities and any subscripts detailing the time or Newton step are dropped for the sake of clearness. The latter implies that all quantities are evaluated at the current Newton-iteration at the current time step t_{n+1} unless indicated by $(\bullet)_n$ referring to the last (converged) time step.

3.4.1 Kirchhoff-Love Beam Elements Based on Constraint Translation

The Kirchhoff-Love beam formulation based on constraint translation as introduced in Sect. 2.5.2 and detailed in Eq. (2.128) is combined with the rotation parameterization based on a rotation vector as proposed in Sect. 2.3.3.1 and the two different rotation interpolation strategies from Sect. 3.2.3. In Eq. (2.128), the highest occurring derivative of an unknown field is of first order calling for C^0-continuity at the element borders. Therefore, Lagrangian and generalized shape functions are employed.

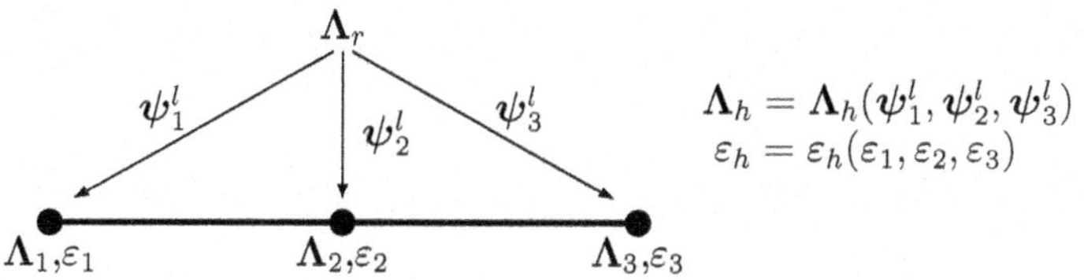

Fig. 3.2 Schematic representation of the $\boldsymbol{\psi}^l$-element with 3 nodes: At every node $\boldsymbol{\psi}_i^l$ is computed, which relates $\boldsymbol{\Lambda}_r$ to $\boldsymbol{\Lambda}_i$ and which is employed to interpolate the orthogonal rotation. $\boldsymbol{\Lambda}_i$ and ε_i are stored and updated at every node[1]

The exceptional point $\boldsymbol{r}_h^*$ is not treated explicitly in the following for the sake of conciseness. In all static cases at least one point in every dimension needs to be fixed anyways. In the case of dynamic problems the point needs to be associated with an inertial mass. Since the case of a free dynamic chain was not treated this remains an open problem.

3.4.1.1 Kirchhoff-Love Beam Element Based on Constraint Translation Using Local Rotation Vector

The Kirchhoff-Love beam element presented in this section is based upon Eq. (2.128) and the rotation interpolation strategy using a local rotation vector $\boldsymbol{\psi}^l$ and generalized shape functions as presented in Sect. 3.2.3. Thus, the trial subspace $(\boldsymbol{r}_h^*, \varepsilon_h, \boldsymbol{\Lambda}_h) \in \mathcal{T}_{\boldsymbol{\psi}^l}$ or alternatively $(\boldsymbol{r}_h^*, \varepsilon_h, \boldsymbol{\psi}_h^l) \in \mathcal{T}_{\boldsymbol{\psi}^l}$ and the weighting subspace $(\delta\boldsymbol{r}_h^*, \delta\varepsilon_h, \delta\boldsymbol{\theta}_h) \in \mathcal{W}_{\boldsymbol{\psi}^l}$ are introduced in a Petrov-Galerkin manner. Here, an element with three nodes at $\xi = -1, \xi = 0$, and $\xi = +1$ is presented. Each node features 4 degrees of freedom, with the exception of the node at $\boldsymbol{r}^*$. Higher order Lagrangian polynomials could be easily employed to obtain elements with an increased number of nodes. This element type will be referred to as $\boldsymbol{\psi}^l$-element and is schematically depicted in Fig. 3.2. The vectors of nodal values of the trail and test functions and the increments $\hat{\mathbf{q}}_{\boldsymbol{\psi}^l}^e$, $\delta\hat{\mathbf{q}}_{\boldsymbol{\psi}^l}^e$, and $\Delta\hat{\mathbf{q}}_{\boldsymbol{\psi}^l}^e$, respectively, are analogously indexed.

Element Residual Vector: The element residual vectors are similar to the ones found for the classic Simo-Reissner beam presented in Sect. 3.3 due to the identical rotation interpolation reading

[1] Figure reprinted with permission from Schulz & Böl (2019). ©2019 by Springer Nature

$$\left(\delta\hat{\mathbf{q}}^e_{\psi^l}\right)^T \mathbf{r}^e_{\text{int},\psi^l} = \int\limits_{\Omega^{\text{el}}_h} \left[\delta\varepsilon_h EA\varepsilon_h + \delta\boldsymbol{\theta}'^T_h \boldsymbol{\Lambda}_h \boldsymbol{M}_h\right] \mathrm{d}s_0, \tag{3.87}$$

$$\left(\delta\hat{\mathbf{q}}^e_{\psi^l}\right)^T \mathbf{r}^e_{\text{kin},\psi^l} = \int\limits_{\Omega^{\text{el}}_h} \left[\delta\boldsymbol{r}^T_h \rho A \ddot{\boldsymbol{r}}_h + \delta\boldsymbol{\theta}^T_h \boldsymbol{\Lambda}_h [\boldsymbol{S}(\boldsymbol{W}_h)\boldsymbol{C}_\rho \boldsymbol{W}_h + \boldsymbol{C}_\rho \boldsymbol{A}_h]\right] \mathrm{d}s_0, \tag{3.88}$$

$$\left(\delta\hat{\mathbf{q}}^e_{\psi^l}\right)^T \mathbf{r}^e_{\text{ext},\psi^l} = \int\limits_{\Omega^{\text{el}}_h} \left[\delta\boldsymbol{r}^T_h \boldsymbol{f} + \delta\boldsymbol{\theta}^T_h \boldsymbol{l}\right] \mathrm{d}s_0 + [\delta\boldsymbol{r}^T_h \bar{\boldsymbol{n}} + \delta\boldsymbol{\theta}^T_h \bar{\boldsymbol{m}}]_{\Gamma_\sigma}. \tag{3.89}$$

The center-line and its derivatives are defined implicitly, but these relations given in Eqs. (2.120)–(2.124) are not repeated in their discretized from. Consequently, the residual vector contains double integrals complicating the implementation. This topic is treated in Sect. 3.4.4.

Element Stiffness Matrix: The element stiffness matrix of the Kirchhoff-Love beam based on constraint translation due to internal forces is less complex than the one found for the classic Simo-Reissner beam due to the simplified translational contributions yielding

$$\left(\delta\hat{\mathbf{q}}^e_{\psi^l}\right)^T \mathbf{k}^e_{\text{int},\psi^l} \Delta\hat{\mathbf{q}}^e_{\psi^l} = \int\limits_{\Omega^{\text{el}}_h} \begin{bmatrix} \delta\varepsilon'_h \\ \delta\boldsymbol{\theta}'_h \end{bmatrix}^T \begin{bmatrix} EA & \mathbf{0} & \mathbf{0} \\ \mathbf{0} & \left(\boldsymbol{\Lambda}_h\, \boldsymbol{S}(\,\boldsymbol{M}_h)\, \boldsymbol{\Lambda}^T_h\right) & \boldsymbol{c}_M \end{bmatrix} \begin{bmatrix} \Delta\varepsilon_h \\ \Delta\boldsymbol{\theta}_h \\ \Delta\boldsymbol{\theta}'_h \end{bmatrix} \mathrm{d}s_0. \tag{3.90}$$

In contrast to the simplified stiffness matrix resulting from the linearization of internal forces, the contribution of inertial forces is more complicated due to the implicit definition of the center-line and its derivatives resulting in

$$\left(\delta\hat{\mathbf{q}}^e_{\psi^l}\right)^T \mathbf{k}^e_{\text{kin},\psi^l} \Delta\hat{\mathbf{q}}^e_{\psi^l} = \tag{3.91}$$

$$\int\limits_{\Omega^{\text{el}}_h} \begin{bmatrix} \delta\boldsymbol{r}_h \\ \delta\boldsymbol{\theta}_h \end{bmatrix}^T \begin{bmatrix} \rho A \dfrac{\partial \ddot{\boldsymbol{r}}_h}{\partial \varepsilon_h} & \rho A \dfrac{\partial \ddot{\boldsymbol{r}}_h}{\partial \boldsymbol{r}_h} \dfrac{\partial \boldsymbol{r}_h}{\partial \boldsymbol{\theta}_h} \\ \mathbf{0} & -\boldsymbol{S}(\boldsymbol{\pi}_h) + \boldsymbol{\Lambda}_h \boldsymbol{C}_\rho \dfrac{\partial \boldsymbol{A}_h}{\partial \boldsymbol{\theta}_h} + \boldsymbol{\Lambda}_h \left(\boldsymbol{S}(\boldsymbol{W}_h) - \boldsymbol{S}(\boldsymbol{C}_\rho \boldsymbol{W}_h)\right) \dfrac{\partial \boldsymbol{W}_h}{\partial \boldsymbol{\theta}_h} \end{bmatrix} \begin{bmatrix} \Delta\varepsilon_h \\ \Delta\boldsymbol{\theta}_h \end{bmatrix} \mathrm{d}s_0$$

$$+ \int\limits_{\Omega^{\text{el}}_h} \left(\Delta\delta\boldsymbol{r}^T_h \ddot{\boldsymbol{r}}_h\right) \mathrm{d}s_0.$$

Using the relation $\Delta\boldsymbol{g}_1 = \boldsymbol{S}(\Delta\boldsymbol{\theta})\boldsymbol{g}_1$, it can easily be verified that the linearized center-line variation at the position s_0 is given by

$$\Delta\delta\boldsymbol{r}(s_0)^T_h = \int_{s^*_0}^{s_0} \left(\delta\boldsymbol{\theta}^T_h [(1+\varepsilon_h)\boldsymbol{S}(\Delta\boldsymbol{\theta}_h \times \boldsymbol{g}_{1,h}) + \boldsymbol{S}(\boldsymbol{g}_{1,h})\Delta\varepsilon_h] - \delta\varepsilon_h [\boldsymbol{S}(\boldsymbol{g}_{1,h})\Delta\boldsymbol{\theta}_h]\right) \mathrm{d}s_0. \tag{3.92}$$

Moreover, employing the temporal discretization techniques from Sects. 3.1.2 and 3.1.3 it follows

$$\frac{\partial \ddot{\boldsymbol{r}}_h}{\partial \varepsilon_h}(s_0) = \int\limits_{s_0^*}^{s_0} \boldsymbol{\Lambda} \left[\frac{1-\alpha_m}{\beta \Delta t^2 (1-\alpha_f)} + 2\frac{\gamma}{\beta \Delta t} \boldsymbol{S}(\boldsymbol{W}) + (\boldsymbol{S}(\boldsymbol{A}) + \boldsymbol{S}(\boldsymbol{W})^2) \right] \boldsymbol{E}_1 \mathrm{d}s_0 \tag{3.93}$$

and $\frac{\partial \ddot{\boldsymbol{r}}_h}{\partial \boldsymbol{\theta}_h}$ results similarly using Eqs. (3.85) and (3.86).

Thus, the stiffness matrix also contains double integrals heavily decreasing the sparseness of the global stiffness matrix and increasing the complexity of the solving process of the system of equations, see also Sect. 3.4.4. This problem seems to be a promising field for improvements, which could be addressed in future work. However, as inertia effects can be neglected for the envisioned application in polymer mechanics, the topic is disregarded at present.

Assuming deformation-independent external loads, the element stiffness matrix can be determined to be

$$\left(\delta \hat{\mathbf{q}}^e_{\boldsymbol{\psi}^l}\right)^T \mathbf{k}^e_{\text{ext},\boldsymbol{\psi}^l} \Delta \hat{\mathbf{q}}^e_{\boldsymbol{\psi}^l} = \int\limits_{\Omega_h^{\text{el}}} \left(\Delta \delta \boldsymbol{r}^T \boldsymbol{f}\right) \mathrm{d}s_0 + [\Delta \delta \boldsymbol{r}^T \overline{\boldsymbol{n}}]_{\Gamma_\sigma} \tag{3.94}$$

due to external loads. It is obvious that every external work, even though itself is independent of the deformation, contributes to the stiffness matrix.

3.4.1.2 Kirchhoff-Love Beam Element Based on Constraint Translation Using Smallest-Rotation Scheme

The Kirchhoff-Love beam element presented in this section is based upon Eq. (2.128) and the rotation interpolation strategy using a split into smallest-rotation and torsional rotation from Sect. 2.3.3.2. Thus, the trial subspace $(\boldsymbol{r}^*_h, \varphi^l_h, \boldsymbol{t}_h) \in \mathcal{T}_t$ and the weighting subspace $(\delta \boldsymbol{r}^*_h, \delta \varphi^l_h, \delta \boldsymbol{t}_h) \in \mathcal{W}_t$ are introduced in a Bubnov-Galerkin manner. An element with three nodes at $\xi = -1$, $\xi = 0$, and $\xi = +1$ is presented. Each node features 4 degrees of freedom, with the exception of the node at $\boldsymbol{r}^*$. Higher order Lagrangian polynomials could be easily employed to obtain elements with an increased number of nodes. This element type will be referred to as $\boldsymbol{t}$-element and is schematically depicted in Fig. 3.3. Note that as with the $\boldsymbol{\varphi}^l$-element the stored nodal quantities are identical but the update procedures differ.

The element residual vectors $\mathbf{r}^e_{\text{int},\boldsymbol{t}}$, $\mathbf{r}^e_{\text{kin},\boldsymbol{t}}$, and $\mathbf{r}^e_{\text{ext},\boldsymbol{t}}$ as well as the element stiffness matrices $\mathbf{k}^e_{\text{int},\boldsymbol{t}}$, $\mathbf{k}^e_{\text{kin},\boldsymbol{t}}$, and $\mathbf{k}^e_{\text{ext},\boldsymbol{t}}$ are almost identical to the ones from the alternative Kirchhoff-Love beam presented in Sect. 3.4.1.1. Besides the interpolation of $\boldsymbol{\Lambda}_h$ only the variations and increments of ε_h and $\boldsymbol{\theta}_h$ need to be given in terms of the variations and increments of $\boldsymbol{t}_h$ and φ_h. These relations are derived in Sect. 2.3.3.2 and are not repeated here in a discretized form.

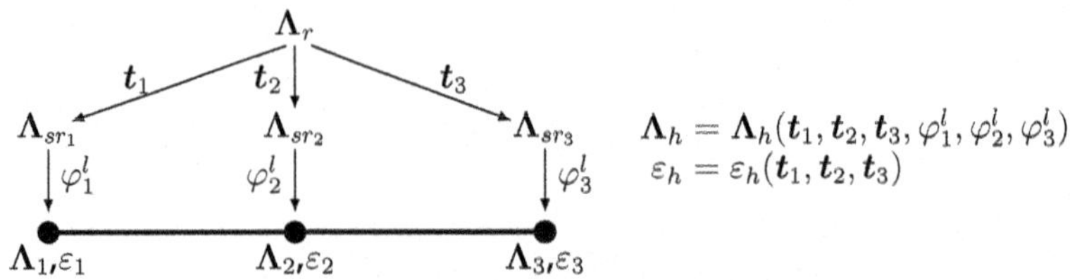

Fig. 3.3 Schematic representation of the $\boldsymbol{t}$-element with 3 nodes: A smallest-rotation part described by $\boldsymbol{\Lambda}_{sr_i}$ and a torsional part φ_i^l describe the rotation from the element-specific reference $\boldsymbol{\Lambda}_r$ to the nodal rotation tensor. $\boldsymbol{\Lambda}_i$ and ε_i are stored and updated at every node and used to compute $\boldsymbol{t}_i$[2]

3.4.2 Inextensible Kirchhoff-Love Beam Element

In Sect. 2.5.5.1, a new inextensible Kirchhoff-Love beam formulation was presented. Its finite-element formulation is almost identical to the one introduced in Sect. 3.4.1.1 except with all terms related to variations and increments of ε_h removed. Then, the trial subspace $(\boldsymbol{r}_h^*, \boldsymbol{\Lambda}_h) \in \mathcal{T}_\varepsilon$ or $(\boldsymbol{r}_h^*, \boldsymbol{\psi}_h^l) \in \mathcal{T}_{\boldsymbol{\psi}^l}$, respectively, and the weighting subspace $(\delta\boldsymbol{r}_h^*, \delta\boldsymbol{\theta}_h) \in \mathcal{W}_\varepsilon$ are constructed in a Petrov-Galerkin manner. In Fig. 3.4 such an element with three nodes at $\xi = -1$, $\xi = 0$, and $\xi = +1$ is schematically depicted. Each node features 3 degrees of freedom, with the exception of the node at $\boldsymbol{r}^*$. This element type will be referred to as ε-element.

3.4.3 Kirchhoff-Love Beam Elements Based on Weak Enforcement

To obtain a finite-element Kirchhoff-Love beam formulation based on weak enforcement the CJ-element introduced in Sect. 3.3 is extended by contributions due to the multipliers discretized as detailed in Sect. 3.2.3 effectively giving rise to the trial subspace $(\boldsymbol{r}_h, \boldsymbol{\Lambda}_h, \lambda_2, \lambda_3) \in \mathcal{T}_h$ and the weighting subspace $(\delta\boldsymbol{r}_h, \delta\boldsymbol{\theta}_h, \delta\lambda_2, \delta\lambda_3) \in \mathcal{W}_h$.

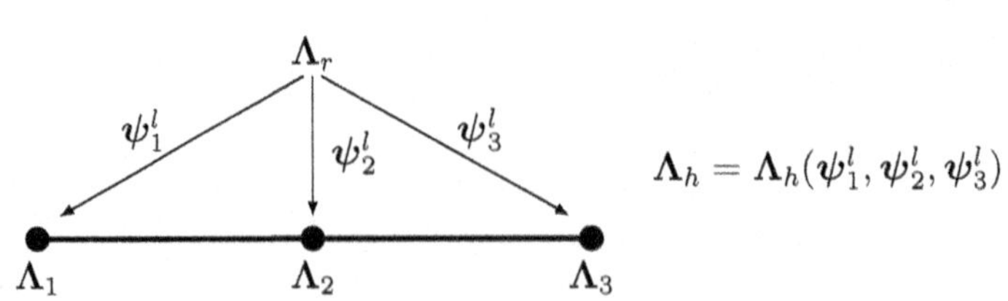

Fig. 3.4 Schematic representation of the ε-element with 3 nodes: At every node $\boldsymbol{\psi}_i^l$ is computed, which relates $\boldsymbol{\Lambda}_r$ to $\boldsymbol{\Lambda}_i$ and which is employed to interpolate the orthogonal rotation. Only $\boldsymbol{\Lambda}_i$ needs to be stored and updated at every node

[2] Figure reprinted with permission from Schulz & Böl (2019). ©2019 by Springer Nature

Thus, the residual vector of the CJ-element is extended by

$$\left(\delta\hat{\boldsymbol{\lambda}}^{e}\right)^{T}\mathbf{r}^{e}_{\text{weak}}=\int\limits_{\Omega_h^{\text{el}}}\left[\delta\lambda_{2,h}(\boldsymbol{g}_{2,h}^{T}\boldsymbol{r}'_{h})+\delta\lambda_{3,h}(\boldsymbol{g}_{3,h}^{T}\boldsymbol{r}'_{h})+\lambda_{2,h}\delta(\boldsymbol{g}_{2,h}^{T}\boldsymbol{r}'_{h})+\lambda_{3,h}\delta(\boldsymbol{g}_{3,h}^{T}\boldsymbol{r}'_{h})\right]\mathrm{d}s_{0}. \tag{3.95}$$

The contribution to the stiffness matrix follows trivially and is not shown here. This element is referred to as the λ-element. It should be noted that there are other ways to obtain weak formulations. For example collocation methods can be employed in combination with Hermite shape functions as done by Meier et al. (2017). But a closer inspection of different enforcement techniques does not lie in the scope of this work.

3.4.4 Nested Assembly Processes

Owing to the indirect definition of kinematic quantities in the Kirchhoff-Love beam finite-element formulations presented in the previous sections, additional assembly processes are required. These nested assembly processes will be explained exemplary on the basis of the $\boldsymbol{\varphi}^{l}$-element. Other assembly procedures using e.g. $\boldsymbol{t}$ and φ^{l} emerge analogously and are not explicitly shown. In contrast to previous definitions the nodal values of variations and increments are removed.

Merging the previously stated results and definitions, the discretized center-line variation $\delta\boldsymbol{r}_h$ is formulated as

$$\delta\boldsymbol{r}_h(s_0)^T=\delta\mathbf{Q}^T\boldsymbol{T}_1(s_0) \tag{3.96}$$

with

$$\boldsymbol{T}_1(s_0)=\mathop{\mathsf{A}}_{s_0^*}^{s_0}\int_{\Omega_h^{\text{el}}}\left(\mathbf{L}^e(1+\varepsilon_h)\boldsymbol{S}(\boldsymbol{g}_{1,h})+\mathbf{L}^e\boldsymbol{g}_{1,h}^T\right)\mathrm{d}s_0. \tag{3.97}$$

Herein, $\boldsymbol{T}_1(s_0)$ results from the discretization of $\delta\boldsymbol{r}(s_0)$ and is introduced for enabling a compact definition. Practically, $\boldsymbol{T}_1(s_0)$ needs to be available at every Gauss point but due to its additive nature the computational effort increases only linearly with the number of nodal unknowns. Using the new compact formulation the discrete residual vector of external forces

$$\mathbf{R}_{\text{ext}}=\mathop{\mathsf{A}}_{e=1}^{n_{\text{ele}}}\int_{\Omega_h^{\text{el}}}\left(\mathbf{L}^e\tilde{\boldsymbol{m}}+\boldsymbol{T}_1\tilde{\boldsymbol{f}}\right)\mathrm{d}s_0+[\mathbf{L}^e\overline{\boldsymbol{m}}+\boldsymbol{T}_1\overline{\boldsymbol{f}}]_{\Gamma_\sigma} \tag{3.98}$$

follows. The required nested assembly procedure has become obvious.

As a second example, the linearization of $\delta\boldsymbol{r}_h$ multiplied by an arbitrary external point- or line-load $\boldsymbol{f}_e$

$$\Delta \delta \boldsymbol{r}_h(s_0)^T \boldsymbol{f}_e = \delta \mathbf{Q}^T \boldsymbol{T}_2(s_0, \boldsymbol{f}_e) \Delta \mathbf{Q} \tag{3.99}$$

with

$$\boldsymbol{T}_2(s_0, \boldsymbol{f}_e) = \mathop{\mathsf{A}}_{s_0^*}^{s_0} \int_{\Omega_h^{\mathrm{el}}} \left(\mathbf{L}^e [(1+\varepsilon_h) \boldsymbol{S}(\boldsymbol{f}_e) \boldsymbol{S}(\boldsymbol{g}_{1,h}) \tilde{\mathbf{L}}^e - \boldsymbol{S}(\boldsymbol{f}_e) \boldsymbol{g}_{1,h} \mathbf{L}^e] - \mathbf{L}^e \boldsymbol{f}_e^T [\boldsymbol{S}(\boldsymbol{g}_{1,h}) \tilde{\mathbf{L}}^e] \right) \mathrm{d}s_0, \tag{3.100}$$

is treated. The global stiffness matrix for deformation-independent loads then reads

$$\mathbf{K}_{\mathrm{ext}} = \mathop{\mathsf{A}}_{e=1}^{n_{\mathrm{ele}}} \int_{\Omega_h^{\mathrm{el}}} \left(\boldsymbol{T}_2(s_0, \bar{\boldsymbol{f}}) \right) \mathrm{d}s_0 + [\boldsymbol{T}_2(s_0, \bar{\boldsymbol{f}})]_{\Gamma_\sigma}. \tag{3.101}$$

Again, as with $\boldsymbol{T}_1$, $\boldsymbol{T}_2$ needs to be available at every Gauss point. Therefore, the amount of the non-zero elements of $\boldsymbol{T}_1(s_0)$ and $\boldsymbol{T}_2(s_0, \boldsymbol{f}_e)$ depend on the total number of elements that are required to discretize the beam in the range from s_0^* to s_0. Also, in Eqs. (3.97) and (3.100) the integral does not necessarily need to be solved over the entire element domain depending on whether s_0 falls within an element or on an element border. A single external point load at the end of the beam can yield a full vector $\mathbf{R}_{\mathrm{ext}}$, whereas $\mathbf{K}_{\mathrm{geo}}$ remains sparse for all load cases.

Upon closer inspection of Eq. (3.91) it can easily be seen that the stiffness matrix due to inertial forces is completely filled since, due to the multiplication of the implicitly defined quantities, all unknown nodal values are connected. As mentioned before, this could be improved in future work but does not impede the current application in polymer mechanics where inertia effects are neglected.

3.4.5 *Dirichlet Boundary Conditions and Joints*

As already mentioned in Sect. 2.5.2.3, boundary conditions need special attention due to the implicit definition of the center-line. In addition, joints are treated in this section for both the $\boldsymbol{\psi}^l$-element and the $\boldsymbol{t}$-element.

Boundary conditions and joints for the $\boldsymbol{\psi}^l$-element: Due to the implicit nature of the beam formulation, in static computations each translational dimension needs to be limited by a given Dirichlet boundary condition at an arbitrary point of the beam. The easiest realization of this requirement is the definition of $\boldsymbol{r}^*$, which can be understood as a simple support and is adapted throughout this work. Additional boundary conditions require the employment of Lagrangian multipliers as was already addressed in Sect. 2.5.2.3. Consequently, the presence of such boundary conditions leads to additional contributions to the residual vector and the stiffness matrix. These contributions are not repeated here in their discretized form. In contrast Dirichlet boundary conditions for rotational degrees of freedom do not require additional multipliers. It suffices to set the rotation increments to zero and reduce the dimension of the stiffness matrix accordingly. Inhomogeneous boundary conditions are not treated here but could be realized under consideration of the special treatment of rotations, see Meier et al. (2017).

A simple moment-free joint is easily realized by only restricting the axial strain to be equal and adjust the assembly process accordingly by removing the respective entries in all global quantities, see e.g. Zienkiewicz & Taylor (2005). A rigid joint requires a little more attention than when encountered in elements that do not involve rotational degrees of freedom. Here, in addition to the axial strain, a relative rotation of the connected nodes has to be suppressed. It can easily be seen that this is the case when the multiplicative rotation increments $\Delta\boldsymbol{\theta}$ of both nodes are set equal. For example, Dirichlet boundary conditions of a fixed end at $s_0 = 0$ can be easily enforced for the $\boldsymbol{\psi}^l$-element by setting

$$\Delta\varepsilon = 0 \text{ and } \Delta\boldsymbol{\theta} = 0 \text{ at } s_0 = 0 \tag{3.102}$$

due to the fact that $\boldsymbol{\Lambda}(s_0 = 0) = \boldsymbol{\Lambda}_0(s_0 = 0)$ is required. Consequently, the joint is realized by adjusting the global system accordingly. Using the material counterpart $\Delta\boldsymbol{\Theta}$ is not possible in the formulation of the joint since the fixed relative orientation is a material quantity defined on the reference. Also, the usage of the additive rotation increment complicates the procedure and is not treated here.

Boundary conditions and joints for the $\boldsymbol{t}$-element: Defining supports and translational Dirichlet boundary conditions involves $\boldsymbol{r}^*$ and Lagrangian multipliers as was the case for the $\boldsymbol{\psi}^l$-element, see the preceding paragraph. In the case of a fixed cross-section at some point along the beam it is required that

$$\boldsymbol{g}_{02}^T \Delta\boldsymbol{t} = \boldsymbol{g}_{03}^T \Delta\boldsymbol{t} = \Delta\varphi = 0. \tag{3.103}$$

Thus, the boundary conditions need to be expressed in terms of $\boldsymbol{g}_{0i}$ and a transformation on element level is required. It should be noted that not all three coordinates of the tangent vector can be described since one value (perpendicular to the cross-section) describes the axial stretch. Inhomogeneous boundary conditions are not treated here.

To form a rotational connection between nodes of different $\boldsymbol{t}$-elements is rather complicated and even requires the introduction of an additional unknown, see Meier et al. (2017), which is not repeated here. Alternatively, it could be reverted to a $\boldsymbol{\psi}^l$-element using the transformation rules stated in Sect. 2.3.3.2 since the realization of joints are easier with that type of element. This could be then referred to as a hybrid approach.

3.5 Requirements of Beam Formulations

In this section, the usual requirements on finite-element (beam) formulations are stated on the one hand, and the newly presented beam formulations from Sect. 3.4 are investigated with regard to these requirements on the other hand.

3.5.1 *Differentiability of Spatially Discretized Fields*

As already detailed in Sect. 3.2.1, it is required that the shape functions are C^{m-1}-continuous if the highest occurring derivative in the weak formulation is of order m. This standard requirement is met in all beam formulations presented in this work.

In contact mechanics, the need for a continuously defined tangent field at the element boarders often leads to the employment of Hermite shape functions. The present contributions, where the tangent field is C^0-continuously represented using Lagrangian shape functions, might be a valuable addition, which could not be tested in the scope of this work.

3.5.2 *Objectivity and Path-Independence*

The problem of ensuring objectivity and path-independence in finite-element formulations is significantly more distinct in beam formulations involving rotational degrees of freedom due to the non-linearity of the configuration space.

In the scope of this work, objectivity is understood as the invariance of spatial quantities under rigid body motion (of the current configuration). To this effect, $(\bullet)^+$ denotes a rigidly transformed (spatial) quantity such that

$$\boldsymbol{r}^+ = \boldsymbol{\Lambda}_R(\boldsymbol{r} + \boldsymbol{r}_R) \text{ and } \boldsymbol{\Lambda}^+ = \boldsymbol{\Lambda}_R\boldsymbol{\Lambda} \tag{3.104}$$

for the strain measures to be objective, where $\boldsymbol{\Lambda}'_R = 0$ and $\boldsymbol{r}'_R = 0$ are true for the rigid rotation and the rigid translation, respectively. It should be noted that a flawed interpolation strategy can render an objective quantity into a discretized quantity that is actually not objective, see Crisfield & Jelenić (1999). The objectivity of the known beam formulations presented in Sects. 2.4 and 2.5.3 has been verified elsewhere, namely in Crisfield & Jelenić (1999) and in Meier et al. (2014), respectively. Thus it remains the study of the objectivity of the $\boldsymbol{\psi}^l$-element and the $\boldsymbol{t}$-element.

When intending on theoretically demonstrating the objectivity, the discretized strain measures are studied. But since the CJ-element and the $\boldsymbol{\psi}^l$-element (and the ε-element) share identical rotation parameterization and therefore the identical rotational strain measure $\boldsymbol{\Omega}$, it only remains to study the objectivity of ε for the $\boldsymbol{\psi}^l$-element, which is done here. Based on Eq. (2.120) the rigidly transformed, implicitly defined, and discretized position vector reads

$$
\begin{aligned}
\boldsymbol{r}_h^+ &= (\boldsymbol{r}^*)^+ + \int_{s_0^*}^{s_0} (1+\varepsilon_h^+)\left(\exp\left[\boldsymbol{S}(\boldsymbol{\psi}_h^l)\right]\boldsymbol{\Lambda}_r\boldsymbol{e}_1\right)^+ \mathrm{d}s_0 + \boldsymbol{\Lambda}_R\boldsymbol{r}_R \\
&= \boldsymbol{\Lambda}_R\boldsymbol{r}^* + \int_{s_0^*}^{s_0} (1+\varepsilon)\exp\left[\boldsymbol{S}(\boldsymbol{\Lambda}_R\boldsymbol{\psi}_h^l)\right]\boldsymbol{\Lambda}_R\boldsymbol{\Lambda}_r\boldsymbol{e}_1\mathrm{d}s_0 + \boldsymbol{\Lambda}_R\boldsymbol{r}_R \\
&= \boldsymbol{\Lambda}_R(\boldsymbol{r}_h + \boldsymbol{r}_R),
\end{aligned} \tag{3.105}
$$

effectively meeting the requirements of objectivity formulated in Eq. (3.104). Furthermore, due to $\boldsymbol{r}'_R = 0$

$$
||\boldsymbol{r}'^+|| = (\boldsymbol{r}'^T\boldsymbol{\Lambda}_R^T\boldsymbol{\Lambda}_R\boldsymbol{r}')^{1/2} \tag{3.106}
$$

and thus

$$
\varepsilon^+ = ||\boldsymbol{r}'^+|| - 1 = ||\boldsymbol{r}'|| - 1 = \varepsilon, \tag{3.107}
$$

which concludes the study of objectivity for the $\boldsymbol{\psi}^l$-element. A numerical conformation is supplemented in Sect. 3.6.1.

In Sect. 2.3.3.2 it can be found how the curvature vector and thus the rotational strain measure can be expressed in terms of the primary variables of the $\boldsymbol{t}$-element. The study of the objectivity of those quantities is started by investigating the transformation behavior of the discretized rotation tensor reading

$$
\boldsymbol{\Lambda}_h^+ = \boldsymbol{R}(\boldsymbol{g}_{1,h}^+, \varphi^+)\boldsymbol{\Lambda}_{sr}(\boldsymbol{g}_{0,h}^+, \boldsymbol{g}_{1,h}^+)\boldsymbol{\Lambda}_R\boldsymbol{\Lambda}_r. \tag{3.108}
$$

It can easily be verified that

$$
\boldsymbol{R}(\boldsymbol{g}_{1,h}^+, \varphi^+) = \boldsymbol{\Lambda}_R\boldsymbol{R}(\boldsymbol{g}_{1,h}, \varphi)\boldsymbol{\Lambda}_R^T \text{ and } \boldsymbol{\Lambda}_{sr}(\boldsymbol{g}_{0,h}^+, \boldsymbol{g}_{1,h}^+) = \boldsymbol{\Lambda}_R\boldsymbol{\Lambda}_{sr}(\boldsymbol{g}_{0,h}, \boldsymbol{g}_{1,h})\boldsymbol{\Lambda}_R^T \tag{3.109}
$$

and thus $\boldsymbol{\Lambda}_h^+ = \boldsymbol{\Lambda}_R\boldsymbol{\Lambda}_h$ as should be. The objectivity of the strain measure follows straight-forward. Analogously to the results of the $\boldsymbol{\psi}^l$-element, the rigidly transformed center-line vector reads $\boldsymbol{r}_h^+ = \boldsymbol{\Lambda}_R(\boldsymbol{r}_h + \boldsymbol{r}_R)$ as well based on the last results, thereby conforming the objectivity of ε. A numerical study is supplemented in Sect. 3.6.1.

A rigorous study of path-independence is more difficult to realize. It can merely be stated that the interpolation in both elements does not use any history values of interpolated quantities and thus they can be expected to be path-independent (Crisfield & Jelenić, 1999). An appropriate numerical study is presented in Sect. 3.6.2.

3.5.3 Numerical Locking

Membrane locking, a special type of locking (Fried & Malkus, 1975; Stolarski & Belytschko, 1982), has already been encountered in various geometrically exact

Kirchhoff-Love beams (Armero & Valverde, 2012a, b; Meier et al., 2014, 2015). It leads to errors in numerical computations and is the result of an unfavorable coupling of kinematic quantities, which are used in the definition of both, the axial strain measure and the curved geometry. This coupling prevents beam element formulations from correctly representing inextensibility of curved structures. Therefore, the more distinct the difference between the axial and bending stiffness is, the higher the numerical errors are. Coherently, Meier et al. (2015) identified the slenderness $\rho := L/r$ as an indication value for locking phenomena.

The classical beam formulations from Jelenić & Crisfield (1999) and Simo (1985) and the first geometrically exact Kirchhoff-Love beam from Meier et al. (2014) suffer from locking phenomena, which can be alleviated using techniques applied at the level of finite-element technology, e.g. reduced and selective integration (Noor & Peters, 1981a; Prathap, 1985; Prathap & Naganarayana, 1990) and penalty relaxation and stabilization techniques (Lyly et al., 1993; Tessler & Spiridigliozzi, 1986). Other techniques start at the level of weak form derivation, e.g. the employment of assumed strains based on the Hu-Washizu functional (Choi & Lim, 1995; Karamanlidis & Jasti, 1987; Lee & Sin, 1993) or of assumed stresses based on the Hellinger-Reissner functional (Cannarozzi & Molari, 2008; Noor & Peters, 1981a). Membrane locking can also be removed by changing the configuration space as done by (Sonneville et al., 2014a, b) who formulated a beam on the special Euclidean group. The effectiveness of this technique will also be seen in the following treatment, where it is theoretically examined if membrane locking is to be expected in the new elements presented in this work.

For the $\boldsymbol{\psi}^l$-element and the ε-element, membrane locking is expected to vanish due to the choice of configuration space. The primary variables representing the rotational and translational degrees of freedom and the respective axial and bending and torsional strain measures are completely decoupled. For a mathematically approach it can be investigated if and under which conditions a constant value of $||\boldsymbol{r}'_h(\xi)|| = 1$ is numerically sustained. Applying this rationale leads to the insight that

$$||\boldsymbol{r}'_h(\xi)|| = ||(1+\varepsilon_h)\boldsymbol{g}_{1,h}|| = ||(1+\sum_{i=1}^{3} L_i(\xi)\varepsilon_i)\boldsymbol{\Lambda}_r \exp\left[\boldsymbol{S}\left(\sum_{i=1}^{3} L_i(\xi)\boldsymbol{\psi}_i\right)\right]\boldsymbol{E}_1|| \tag{3.110}$$

holds, where

$$||\boldsymbol{\Lambda}_r \exp\left[\boldsymbol{S}\left(\sum_{i=1}^{3} L_i(\xi)\boldsymbol{\psi}_i\right)\right]\boldsymbol{E}_1|| = 1 \tag{3.111}$$

is true for $\xi \in [-1, +1]$ and

$$||(1+\sum_{i=1}^{3} L_i(\xi)\varepsilon_i)|| = 1 \tag{3.112}$$

always exists for $\varepsilon_1 = \varepsilon_2 = \varepsilon_3 = 0$. Thus, an inextensible geometry can be represented correctly.

Changing the primary fields, alters the result for the $\boldsymbol{t}$-element, where the same approach leads to

$$||\boldsymbol{r}_h'(\xi)|| = ||L_1(\xi)\boldsymbol{t}_1 + L_2(\xi)\boldsymbol{t}_2 + L_3(\xi)\boldsymbol{t}_3||, \tag{3.113}$$

which reveals that $||\boldsymbol{r}_h'(\xi)|| = 1$ is true for $\xi \in [-1, +1]$ only if $\boldsymbol{t}_1 = \boldsymbol{t}_2 = \boldsymbol{t}_3$ and $||\boldsymbol{t}_1|| = ||\boldsymbol{t}_2|| = ||\boldsymbol{t}_3|| = 1$. The only solution for this problem is a straight line indicating the possibility of membrane locking. To further investigate the beam formulation with regard to membrane locking a small computation based on Hughes (2012) can be performed, similar to an investigation done in Meier et al. (2015). Hughes (2012) introduced the constraint ratios for the continuous and the discretized problem reading

$$r = \frac{n_{\text{eq}}}{n_c} \quad \text{and} \quad r_h = \lim_{n_{\text{el}} \to \infty} \frac{n_{\text{node}} \cdot n_{\text{DOF}}}{n_{c,h}}, \tag{3.114}$$

respectively. Therein, n_{eq} signifies the number of differential equations describing the components of $\boldsymbol{r}$ and n_c denote the number of constraint equations. Furthermore, n_{node}, n_{DOF}, and $n_{c,h}$ define the number of nodes, degrees of freedom per node, and constraints of the discretized problem, respectively. In the continuous version of the $\boldsymbol{t}$-element, $n_{\text{eq}} = 4$ and $n_c = 1$ such that $r = 4$. For the three-node $\boldsymbol{t}$-element, $n_{\text{node}} = 2(n_{\text{el}} + 1) - 6$, $n_{\text{DOF}} = 4$, and $n_{c,h} = 4 \cdot n_{\text{el}}$ such that $r_h = 2$. Thus, it is established that $r > r_h$. According to Hughes (2012), this finding strengthens the indication of locking, which will be further investigated with the aid of numerical examples in Sect. 3.6.3.

3.5.4 *Conservation Properties*

It is clear that any weak form of a mechanical problem needs to adhere to the conservation of linear momentum, angular momentum, and energy. Expectably, the same holds for the spatially discretized formulations. These requirements can be used to formulate general preconditions for beam formulations as briefly presented in the following.

Under the absence of external forces, including those resulting from bearings or other boundary conditions, the linear momentum is conserved if a virtual motion corresponding to a rigid body translation defined by

$$\mathcal{W}_h \ni \delta\boldsymbol{\theta} \equiv 0 \text{ and } \mathcal{W}_h \ni \delta\boldsymbol{r} \equiv \boldsymbol{u}_{\text{c}} \text{ with } \boldsymbol{u}_{\text{c}}' = \boldsymbol{0} \tag{3.115}$$

can be exactly represented by the discretized element formulation. Analogously, angular momentum is conserved if a virtual rigid body rotation defined by

$$\mathcal{W}_h \ni \delta\boldsymbol{\theta} \equiv \boldsymbol{w}_c \text{ and } \mathcal{W}_h \ni \delta\boldsymbol{r} \equiv \boldsymbol{w}_c \times \boldsymbol{r}_h \text{ with } \boldsymbol{w}'_c = \boldsymbol{0} \tag{3.116}$$

can be exactly represented. The requirements can also be formulated differently: The weighing functions defined in Eqs. (3.115) and (3.116) need to be contained within the discrete weighting space of the finite element formulation.

In the study of the conversion of energy, the sum of kinetic and internal energy, see Sect. 2.5.2.3, is considered. It is required that under the absence of external (mechanical) power the sum of all rates of change of energy remains zero. The corresponding virtual motion

$$\mathcal{W}_h \ni \delta\boldsymbol{\theta} \equiv \boldsymbol{w}_h \text{ and } \mathcal{W}_h \ni \delta\dot{\boldsymbol{r}}_h \tag{3.117}$$

needs to be contained within the respective discretized weighting space. In the following it is verified that the requirements formulated in Eqs. (3.115) and (3.116) are met and that these variations are in fact contained within $\mathcal{W}_h$. The problem of conserving energy involves studying and refining time discretization, which exceeds the scope of this thesis.

Conservation properties of the $\boldsymbol{\psi}^l$-element: Following Eq. (3.115), $\delta r^* = \boldsymbol{u}_c$ with $\boldsymbol{u}'_c = 0$, $\varepsilon = 0$ and $\delta\boldsymbol{\theta} = 0$ are first plugged into Eq. (2.121) and the obtained result is plugged into Eq. (2.128) yielding

$$\delta\boldsymbol{r} = \boldsymbol{u}_c \tag{3.118}$$

and thus

$$\dot{\boldsymbol{L}} = \int_0^{L_0} \left(-\boldsymbol{u}_c^T \boldsymbol{f}\right) \mathrm{d}s_0 + [\boldsymbol{u}_c^T \overline{\boldsymbol{n}}]_{\Gamma_\sigma}, \tag{3.119}$$

respectively. Therein, $\boldsymbol{L}$ depicts the linear momentum vector and $\boldsymbol{f}_\rho = 0$. It can be seen that for the chosen variations, $\boldsymbol{L}$ is only changed by external loads or boundary conditions. This finding means that to preserve linear momentum in the present implicit formulation, Eq. (3.115) needs to hold, which can be seen to be true for the $\boldsymbol{\psi}^l$-element without further investigation.

Turning towards the conservation of angular momentum and following Eq. (3.116), $\delta\boldsymbol{\theta} = \boldsymbol{w}_c$, $\delta\boldsymbol{r}* = \boldsymbol{w}_c \times \boldsymbol{r}^*$, and $\delta\varepsilon = 0$ with $\boldsymbol{w}'_c = 0$ are first plugged into Eq. (2.121) and then the resulting relation into Eq. (2.128) yielding

$$\delta\boldsymbol{r} = \boldsymbol{w}_c \times \boldsymbol{r}^* + \boldsymbol{w}_c \times (\boldsymbol{r} - \boldsymbol{r}^*) = \boldsymbol{w}_c \times \boldsymbol{r} \tag{3.120}$$

and thus

$$\dot{\boldsymbol{H}} = \int_0^{L_0} \left(-\delta\boldsymbol{r}^T \boldsymbol{f} - \boldsymbol{w}_c \times \boldsymbol{l}\right) \mathrm{d}s_0 + [\boldsymbol{w}_c \times \overline{\boldsymbol{m}} + \delta\boldsymbol{r}^T \overline{\boldsymbol{n}}]_{\Gamma_\sigma}. \tag{3.121}$$

Herein, $\boldsymbol{H}$ depicts the angular momentum vector and $\boldsymbol{m}_\rho = 0$ is evident. It can be concluded that for the chosen variations, $\boldsymbol{H}$ is changed only by external loads or boundary conditions. This finding means that to preserve angular momentum in the present implicit formulation, Eq. (3.116) needs to hold, which requires the underlying discretization of $\delta\boldsymbol{\theta}_h$ to be able to represent a constant value. Conveniently, it was already shown by Jelenić & Crisfield (1999) that the rotation interpolation strategy based on a local rotation vector is indeed able to represent constant curvature and therefore the $\boldsymbol{\psi}^l$-element conserves angular momentum.

Conservation properties of the $\boldsymbol{t}$-element: Based upon the findings of the last section it merely remains to show that a virtual rigid body translation and rotation can be exactly represented using the chosen rotation discretization technique. It is trivial to see that $\delta\boldsymbol{r}^* = \boldsymbol{u}_{c*}$, $\boldsymbol{u}'_{c*} = 0$, $\delta\boldsymbol{t} = \boldsymbol{u}_c$, $\boldsymbol{u}'_c = 0$, and $\varphi^l = 0$ represent such a rigid body translation. But, if the chosen rotation parameterization is actually able to exactly model a constant curvature is not a trivial question. Fortunately, this question was already resolved by Meier et al. (2017) with the interesting result that only a Bubnov-Galerkin variant is actually able to express constant curvature, which is why this approach was chosen in Sect. 3.4.1.2. Meier et al. (2017) were able to use the rotation parameterization scheme of the $\boldsymbol{t}$-element in a Petrov-Galerkin method by omitting certain terms. But as theoretically expected, the latter version did not conserve angular momentum, which is why this variant is not further considered.

Conservation Properties of the ε-Element: The ε-element variant also preserves linear and angular momentum, which can be directly inferred from the results for the $\boldsymbol{\psi}^l$-element.

3.5.5 *Optimal Convergence Order*

Theoretically, the L^2-error for an elliptic variational problem of order m evolves following (Strang & Fix, 2008)

$$||e||^2 = \mathcal{O}(h^{k+1} + h^{2(k-m+1)}). \tag{3.122}$$

Herein, h denotes the element length and k depicts the degree of the shape functions, which is two for all new Kirchhoff-Love beam elements. A short study of the $\boldsymbol{\psi}^l$-element reveals that the integral involved in computing $\delta\boldsymbol{r}$ reduces k and a second-order convergence is expected for the first derivative of the center line $\boldsymbol{r}'_h(\xi)$. In contrast to that, for the $\boldsymbol{t}$-element, $\delta\boldsymbol{r}$ is simply equal to $\delta\boldsymbol{t}$. Hence, no additional integration reduces k and a convergence order of three can be expected for $\boldsymbol{r}'_h(\xi)$. But when considering the center-line $\boldsymbol{r}_h(\xi)$ that results from the tangent, it is expected that the involved integral increases the convergence order, i.e. for the center-line of the $\boldsymbol{\psi}^l$-element and $\boldsymbol{t}$-element third and fourth-order convergence is expected, respectively.

To quantitatively evaluate the L^2-error of the center-line and the tangent field the two errors

$$||e||_{r}^{2} = \sqrt{\frac{1}{L_0}\int_0^L ||\boldsymbol{r}_{\text{ref}} - \boldsymbol{r}_h||^2 \mathrm{d}s_0} \tag{3.123}$$

and

$$||e||_{r'}^{2} = \sqrt{\frac{1}{L_0}\int_0^L ||\boldsymbol{r}'_{\text{ref}} - \boldsymbol{r}'_h||^2 \mathrm{d}s_0} \tag{3.124}$$

are introduced. Therein, the index $(\bullet)_{\text{ref}}$ denotes an either analytically or numerically obtained reference solution, which will be stated in the respective numerical example.

3.6 Numerical Examples

In this section, first, numerical examples are provided to confirm or refute basic properties, i.e. objectivity, path-independence, convergence order, and the absence of membrane-locking of the $\boldsymbol{\psi}^l$-element, the $\boldsymbol{t}$-element, and the ε-element. Secondly, the convergence behavior and performance of these elements are studied and compared to the SR-element (Simo, 1985), the CJ-element (Jelenić & Crisfield, 1999), the MSR-element (Meier et al., 2015), the MLV-element (Meier et al., 2017), and the WE-element based on weak enforcement of the Kirchhoff constraints described in Sect. 2.5.4. The MSR-element and the MLV-element enforce the Kirchhoff constraint by constraining the rotation and they use the smallest-rotation and local rotation vector interpolation techniques from Sect. 3.2.3. If not stated otherwise, the WE-element, the MSR-element, and the MLV-element are equipped with the so-called MCS approach (Meier et al., 2015), which is a collocation-type method used to remove locking phenomena. Furthermore, comparisons are focused on the MSR-element since it was identified as superior by Meier et al. (2017) in terms of accuracy and performance to the other element types mentioned except for the WE-element, which showed similar advantageous properties. Thirdly, a dynamic example is presented.

In the following numerical examples, the $\boldsymbol{\psi}^l$-element, the $\boldsymbol{t}$-element, and the ε-element are constructed using three nodes with quadratic Lagrangian interpolation functions and four Gauss points within the numerical integration scheme. The remaining elements mentioned above are constructed in the most correlating fashion. The MSR-element and MLV-element require Hermitian shape functions, for example, and are constructed using two nodes. Therefore, the comparisons are in relation to the degrees of freedom of the different element formulations.

If not stated otherwise, the beams have an initial length $L_0 = 1000$ and a quadratic cross-section with a varying side length r resulting in different slenderness ratios $\rho := L/r$, which was identified by Meier et al. (2015) to be indicative of membrane-locking phenomena. Furthermore, the following constitutive parameters are chosen and kept equal for all examples: $E = 1.0$ and $G = 0.5$ such that $GI_p = EI_2 = EI_3 = r^4/12$

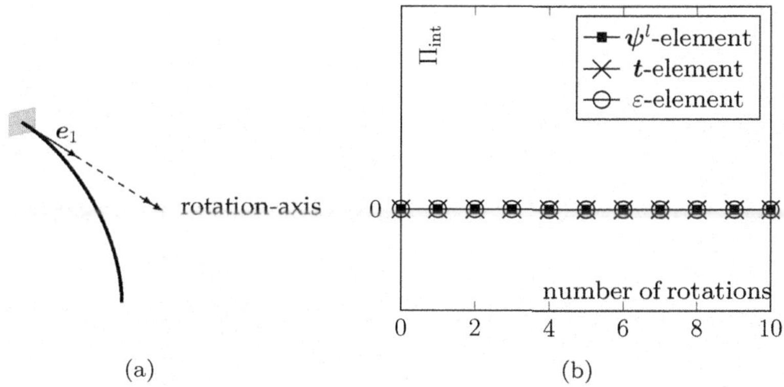

Fig. 3.5 Numerical example to test objectivity of beam formulations: **a** An initially bent beam is mounted such that it can be rotated about the $\boldsymbol{e}_1$-vector, which is realized in a total of 100 load steps until 10 full revelations are completed. **b** The internal energy Π_{int} is computed at all times in dependence of the performed rotations[3]

and $EA = r^2$. Furthermore, all examples are conducted using Dirichlet boundary conditions of a fixed end at $s_0 = 0$.

3.6.1 *Objectivity*

The test case presented in this section is adopted from Meier et al. (2014) to numerically confirm the theoretically expected objectivity of the $\boldsymbol{\theta}^l$-element, the $\boldsymbol{t}$-element, and the ε-element, see Sect. 3.5.2. As shown in Fig. 3.5a, a beam forming a quarter circle is fixed at one end such that it can be rotated about the depicted vector $\boldsymbol{e}_1$. The beam is then rotated about this vector and the internal energy defined in Eq. (2.133) is plotted over the number of rotations about $\boldsymbol{e}_1$, which are realized by multiplying the nodal rotation tensors $\boldsymbol{\Lambda}_i$ by $\exp(\boldsymbol{S}(\Delta\boldsymbol{\theta}))$, where $\Delta\boldsymbol{\theta} = [\Delta\theta_1, 0, 0]_{\boldsymbol{e}_i}$. In more detail, $\Delta\theta_1$ is linearly and step-wise increased to reach 20π in 100 load steps. After every increase the internal energy is determined and the resulting plot is shown in Fig. 3.5b. It can be seen that the rotations do not give rise to an internal energy. Therefore, the beam formulation is considered objective. It can be used for dynamic problem settings, where large rotations occur.

[3] Figure reprinted with permission from Schulz & Böl (2019). ©2019 by Springer Nature

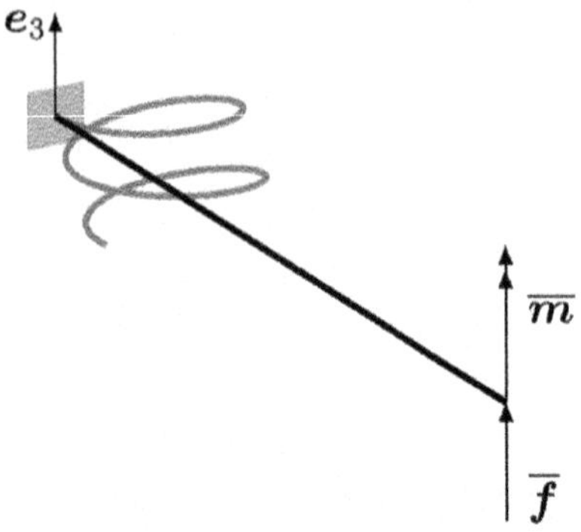

Fig. 3.6 Numerical example to test path-dependency of beam formulations: A straight beam is fixed at one end and is loaded at the other end by a force $\overline{f} = 4EI\pi/L_0^2\, e_3$ and a moment $\overline{m} = 4EI\pi/L_0\, e_3$. The deformed configuration is shown in *gray*[4]

3.6.2 *Path-Independence*

This test case was first presented by Ibrahimbegovic (1997) and has already been applied by e.g. Battini & Pacoste (2002); Zupan & Saje (2003); Meier et al. (2014). Here, it is used to demonstrate path-independence as theoretically studied in Sect. 3.5.2. A beam is fixed at $s_0 = 0$ and at its other end at $s_0 = L_0$ it is loaded with both, a single moment and force defined by $\overline{m}(s_0 = L_0) = (0, 0, 4EI\pi/L_0)^T_{e_i}$ and $\overline{f}(s_0 = L_0) = (0, 0, 4EI\pi/L_0^2)^T_{e_i}$, respectively. The problem set-up is shown in Fig. 3.6 including the deformed configuration as well. To cancel out the membrane-locking topic, a moderate slenderness of $\rho = 100$ is chosen.

To be able to study the path-dependence at least two different load paths need to be defined. The first path is referred to as being the simultaneous (index: sim) path, where both loads, $\overline{m}$ and $\overline{f}$, are linearly increased within the same (pseudo) time interval $t \in [0; 1]$ and $r'_{\mathrm{sim},h}$ signifies the obtained solution. In contrast, the second path is referred to as successive (index: suc) path. As the designation suggests, first the moment is applied within $t \in [0.5; 1]$ and subsequently the force follows within $t \in [0.5; 1]$ yielding the solution $r'_{\mathrm{suc},h}$. Of course, this procedure excludes the ε-element since it definitively results in a different final deformed configuration due to its axial rigidity.

The different load paths result in various deformed configurations for $t < 1$. But at $t = 1$ the deformed configurations of the ψ^l-element and t-element should be identical within numerical accuracy. Thus, to quantitatively evaluate any path-dependence effects, which would cause different final deformed states, the L^2-error (referring to Eq. (3.124))

$$||e||^2_{r'} = \sqrt{\frac{1}{L_0}\int_0^l ||r'_{\mathrm{sim},h} - r'_{\mathrm{suc},h}||^2 \mathrm{d}s_0} \tag{3.125}$$

is computed for different element lengths. The results of that procedure are depicted in Fig. 3.7, where the dependency of $||e||^2_{r'}$ on the element length is illustrated. $||e||^2_{r'}$

[4] Figure reprinted with permission from Schulz & Böl (2019). ©2019 by Springer Nature

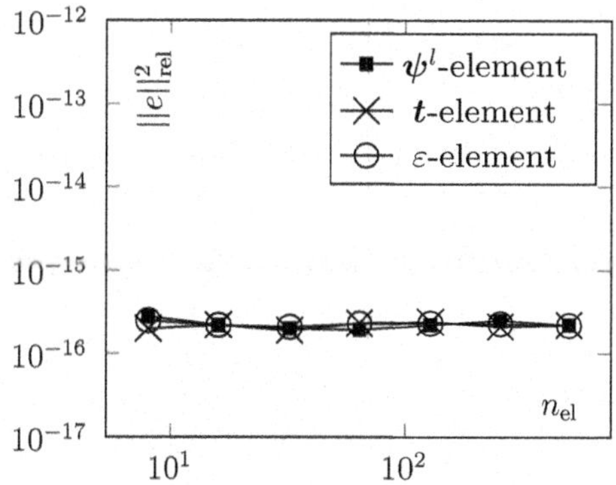

Fig. 3.7 The L^2-error as defined in Eq. (3.125) is shown in dependence on the number of elements according to the test problem depicted in Fig. 3.6 to identify path-dependency effects

assumes values in the range of computer precision and therefore, both beam formulations can be considered as path-independent.

It should also be noted that, surprisingly, in the final configuration at $t = 1$, the tip lies within the negative z-space even though the external load points in the positive z-direction. This result contradicting intuition was also obtained by others (Battini & Pacoste, 2002; Ibrahimbegovic, 1997; Meier et al., 2014; Zupan & Saje, 2003).

3.6.3 *Locking and Convergence Order*

In the following section the theoretical study concerning locking phenomena as presented in Sect. 3.5.3 is supplemented by additional numerical examples. At the same time, to optimally use the examples, it is studied if the convergence order matches the theoretically expected values, see Sect. 3.5.5.

3.6.3.1 Cantilever Loaded by a Moment at Its End

In the following example, a fixed cantilever loaded by a single moment at its end is considered, as has been done by e.g. Bauer et al. (2016) and Simo & Vu-Quoc (1986). The beam is loaded by a single external moment at its end. Here, the external moment is chosen to be $\overline{\boldsymbol{m}}(s_0 = L_0) = (0, 0, EI\pi/(2L_0))^T_{\boldsymbol{e}_i}$, as depicted in Fig. 3.8. For this example an analytical solution is available, see Sect. 2.5.6, and the deformed beam forms a quarter circle. Furthermore, two different slenderness ratios, $\rho = 100$ and $\rho = 10{,}000$, are employed.

For both slenderness ratios the two L^2-errors according to Eqs. (3.123) and (3.124) are computed in dependence of the number of degrees of freedom and shown in Fig. 3.9. For comparison, the L^2-error of the same example using the MSR-element is depicted.

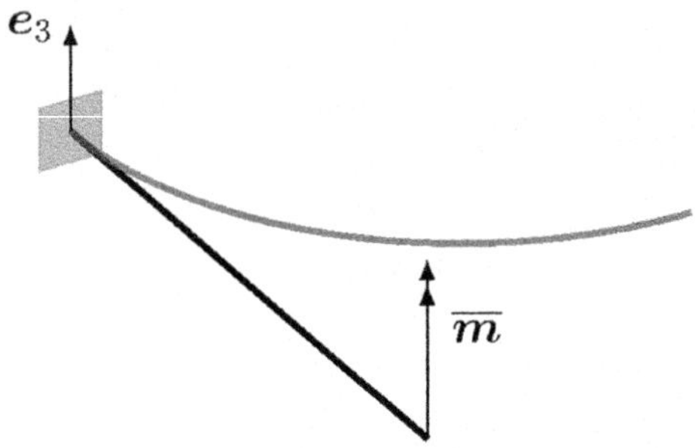

Fig. 3.8 Numerical example to study locking effects and the order of convergence: A straight cantilever is loaded by $\overline{\boldsymbol{m}} = EI\pi/(2L_0)\boldsymbol{e}_3^T$. The deformed configuration is shown in *gray*[5]

For the $\boldsymbol{\psi}^l$-element and ε-element, the L^2-error is around computer precision even when using only a single element. This is due to the employed rotation interpolation scheme, which can exactly represent constant curvature as discussed in Sect. 3.5.4. Consequently a more complex problem is required, which is presented in Sect. 3.6.3.2 where a spring under axial load with non-constant curvature is loaded.

For the $\boldsymbol{t}$-element, the numerical results confirm those obtained theoretically in Sect. 3.5.5. for high enough element numbers the tangent and displacement fields exhibit third and forth-order convergence, respectively. These results hold for both slenderness ratios. It should be pointed out that this result is rather surprising. A closely related variant of the $\boldsymbol{t}$-element was studied by Meier et al. (2015) who found severe locking effects in the same numerical example. However, it was observed that for $\rho = 10{,}000$ the $\boldsymbol{t}$-element does not show a quadratic convergence rate in vicinity of the result but rather heavy fluctuations. Thus, more iteration steps and smaller loading steps are required before convergence is reached as will be further studied in Sects. 3.6.3.2 and 3.6.4, respectively.

In comparison to the MSR-element it can be stated that both new elements are superior in the present example but further studies are required.

3.6.3.2 Twisted Spring Under Axial Load

In the last example it became clear that a more complex beam geometry with non-constant curvature is required to further study the convergence order of the new elements. To this end, an example is taken from Meier et al. (2015), where a single beam with an initially curved geometry is loaded by a single force. As in the previous section, the two different L^2-errors defined in Eqs. (3.123) and (3.124) are plotted in dependence of the number of degrees of freedom n_{dof} for different slenderness ratios.

The initial configuration of the beam in this example is described by the position vector of the center-line

[5] Figure reprinted with permission from Schulz & Böl (2019). ©2019 by Springer Nature

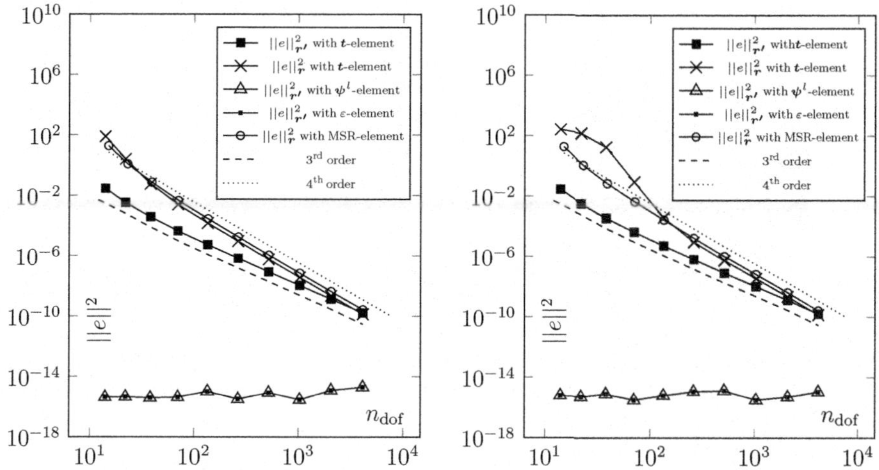

(a) (b)

Fig. 3.9 Numerical example to study locking effects and the order of convergence: The errors according to Eqs. (3.123) and (3.124) are plotted against the number of degrees of freedom for two different slenderness ratios. **a** $\rho = 100$ and **b** $\rho = 10{,}000$

$$\boldsymbol{r}_0(\alpha) = R_0 \begin{pmatrix} [2]\sin\left(\frac{9\pi}{L_0}\alpha\right) \\ \cos\left(\frac{9\pi}{L_0}\alpha\right) - 1 \\ \frac{6}{L_0^2}\alpha^2 \end{pmatrix}_{\boldsymbol{e}_i} \tag{3.126}$$

with

$$R_0 = \frac{L_0}{6\sqrt{\left(\frac{3\pi}{4}\right)^2 + 1} + \frac{27\pi^2}{8}\ln\left(\frac{4}{3\pi} + \sqrt{1 + \left(\frac{4}{3\pi}\right)^2}\right)} \quad \text{and} \quad \alpha \in [0; L_0].$$

The space curve given by Eq. (3.126) forms a twisted spring, as shown in Fig. 3.10, which is loaded in the direction of its axis. The amount of the force changes with the slenderness ratio of the beam. As before, $\rho_1 = 100$ and $\rho_2 = 10{,}000$ are chosen with the corresponding external loads $\overline{\boldsymbol{f}}_{\rho_1}(\alpha = L_0) = (0, 0, 2 \cdot 10^{-1})^T_{\boldsymbol{e}_i}$ and $\overline{\boldsymbol{f}}_{\rho_2}(\alpha = L_0) = (0, 0, 2 \cdot 10^{-9})^T_{\boldsymbol{e}_i}$, respectively.

The beam is divided into elements of equal length, which leads to a total of n_n nodes. But since α is not an arc-length parameter in Eq. (3.126) the values of α at the nodes need to be determined in a numerical manner. First, the normalized tangent

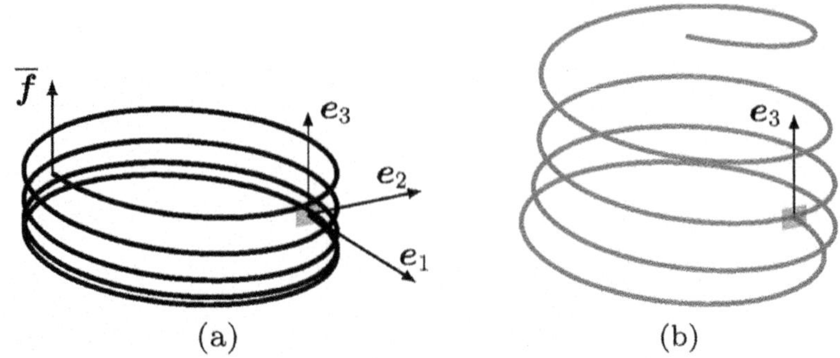

Fig. 3.10 Numerical example to study locking effects and the order of convergence: **a** A curved helix is fixed at one end and loaded at the other end by a single point load in direction of $\boldsymbol{e}_3$ and **b** deformed configuration of the curved helix[6]

$$\boldsymbol{g}_{01} = \frac{\boldsymbol{r}_0'(\alpha)}{|\boldsymbol{r}_0'(\alpha)|} \quad \text{with } \boldsymbol{r}_0'(\alpha) = R_0 \begin{pmatrix} [2]\dfrac{9\pi}{L_0}\cos\left(\dfrac{9\pi}{L_0}\alpha\right) \\ -\dfrac{9\pi}{L_0}\sin\left(\dfrac{9\pi}{L_0}\alpha\right) \\ \dfrac{12}{L_0^2}\alpha \end{pmatrix}_{\boldsymbol{e}_i} \tag{3.127}$$

was computed at each node. Secondly, the nodal rotation tensors $\boldsymbol{\Lambda}_{0,i}$ are obtained by updating the nodal tensors along beam according to

$$\boldsymbol{\Lambda}_{0,i+1} = \boldsymbol{\Lambda}_{0,i}^{\text{rel}}\, \boldsymbol{\Lambda}_{0,i} \qquad \text{for} \quad i = 1, 2, ..., n_n. \tag{3.128}$$

Herein, $\boldsymbol{\Lambda}_{0,i}^{\text{rel}}$ is the tensor that maps $\boldsymbol{g}_{01,i}$ to $\boldsymbol{g}_{01,i+1}$.

The error of the tangent field $||e||_{\boldsymbol{r}'}^2$ for the spring under load is shown in Fig. 3.11 for (a) $\rho_1 = 100$ and (b) $\rho_2 = 10{,}000$. The solution is obtained by using 10,000 WE-elements except for determining the error of the ε-element where 10,000 ε-elements were used to obtain the reference solution.

As expected based on the theoretical considerations presented in Sect. 3.5.2 the $\boldsymbol{\psi}^l$-element and the ε-element strictly follow a second order error reduction and show no dependency on ρ.

The $\boldsymbol{t}$-element, in contrast, is expected to not strictly follow third order convergence rate at least for high slenderness ratios, which is confirmed by the present data. For $\rho_1 = 100$ the $\boldsymbol{t}$-element does not exhibit a constant convergence order but one that can be approximated by a third order reduction. But for $\rho_2 = 10{,}000$, the $\boldsymbol{t}$-element clearly shows convergence to a wrong solution due to numerical locking. The error levels off as can be seen in Fig. 3.11b. Thus it can be noted that the unfavorable constraint ratio of the $\boldsymbol{t}$-element does not impede the computation of simple load-

[6] Figure reprinted with permission from Schulz & Böl (2019). ©2019 by Springer Nature

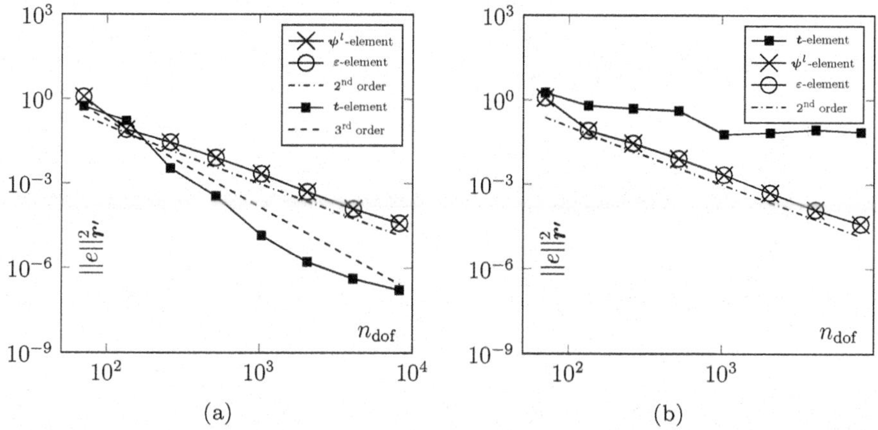

Fig. 3.11 Numerical example to study locking effects and the order of convergence: For the loading case from Fig. 3.10 the relative error as defined in Eq. (3.124) is plotted against the number of degrees of freedom for various element types. Two different slenderness ratios. **a** $\rho = 100$ and **b** $\rho = 10{,}000$ are employed

ing cases as considered in Sect. 3.6.3.1 but for more complicated examples locking effects do occur.

To complete the convergence study of the new elements, $||e||^2_{\boldsymbol{r}}$ is computed for the present examples, compared to the results using elements form literature, and depicted in Fig. 3.12 for (a) $\rho_1 = 100$ and (b) $\rho_2 = 10{,}000$.

Again, as expected, the $\boldsymbol{\psi}^l$-element and the ε-element strictly follow a third order error reduction, which is one order higher than was observed for $||e||^2_{\boldsymbol{r}'}$ due to the smoothing effect of the integral involved. Also, no dependency on ρ is found.

For the $\boldsymbol{t}$-element, the expected fourth order rate is found only for the first case. When increasing the slenderness ratio this finding cannot be confirmed. A second order reduction could be assumed before the error levels off due to numerical locking. The $\boldsymbol{\psi}^l$-element and the MSR-element have similar error levels for both slenderness ratios. For the case of the less slender beam the $\boldsymbol{t}$-element shows better error levels. Equipping the $\boldsymbol{t}$-element with the same method results again in a comparable value of the L^2-error.

3.6.4 Performance

In this section, the performance of different element formulations is compared by counting the iteration steps of the Newton-Raphson scheme. In detail, the adaption scheme from Meier et al. (2017) is employed: Initially, a tenth of the final load is applied. In case the computation does not converge, only half of that load is

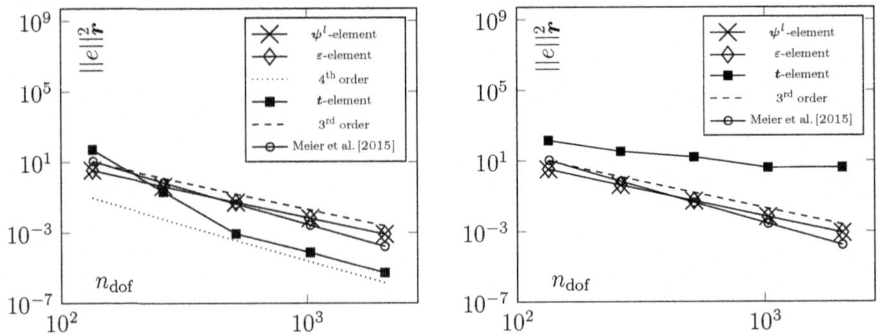

(a) (b)

Fig. 3.12 Numerical example to study locking effects and the order of convergence: The relative error as defined in Eq. (3.124) is shown in dependence of the number of degrees of freedom for different element types for the loading case depicted in Fig. 3.10a. Two different slenderness ratios. **a** $\rho = 100$ and **b** $\rho = 10{,}000$ are employed

applied. But after four successive simulations having converged successfully, the load increment is doubled.

3.6.4.1 Cantilever Loaded by a Moment at Its End

The performance study is started by an example where the beam is only bent by external moments. To this end, as in Sect. 3.6.3, a fixed cantilever is loaded by a moment at its end. Here, $\overline{\boldsymbol{m}}_1 = M\boldsymbol{e}_3$ and $\overline{\boldsymbol{m}}_2 = M\boldsymbol{e}_1$ are applied, where $M = 10$ and $M = 10^{-7}$ for $\rho = 100$ and $\rho = 10{,}000$, respectively. The undeformed and deformed configurations resulting from those loads are shown in Fig. 3.13. The total number of required Newton iterations n_{iter} for both slenderness ratios is plotted against the respective number of elements for different element types and shown in Fig. 3.14.

The performance of the $\boldsymbol{\psi}^l$-element and the ε-element is found to be fluctuating about $n_{\text{iter}} \approx 20$ independently of both, the number of elements and ρ. These two elements are the most efficient elements of the compared formulations for the considered loading case. But it is clear that this overwhelmingly superior behavior is due to the nature of the problem where only external moments are present. An example including not only external moments but also external forces the performance is expected to give rise to different findings. This will be inspected in the next section.

In contrast, the $\boldsymbol{t}$-element shows strong dependency of the element number but not of the slenderness. For small element numbers, high iteration numbers are found. Nonetheless, at higher element numbers it is approximately as performant as the the MSR-element. This is not surprising since it is based on the same rotation interpolation scheme.

Fig. 3.13 Numerical example to study performance: A straight cantilever is loaded by $\overline{m}_1 = M\, \boldsymbol{e}_3$ and $\overline{m}_2 = M\, \boldsymbol{e}_1$ at its end. $M = 10$ and $M = 10^{-7}$ for $\rho = 100$ and $\rho = 10{,}000$, respectively. The deformed configuration is shown in *gray*.[7]

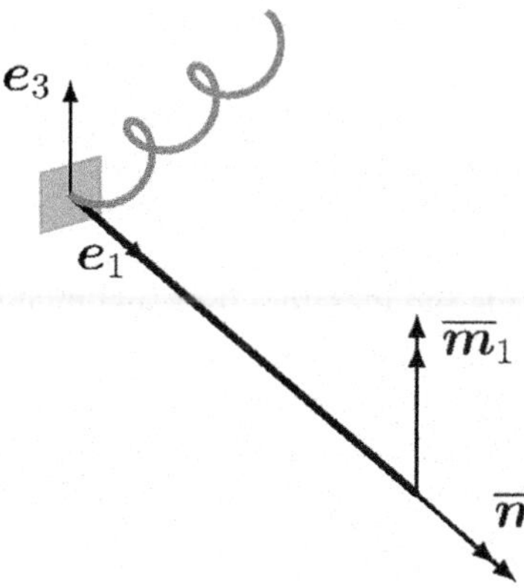

It is pointed out that the beam element formulations devised by Jelenić & Crisfield (1999) and Simo (1985) require $n_{\text{iter}} \approx 12{,}000$ when $\rho = 10{,}000$ and are not included in Fig. 3.14b.

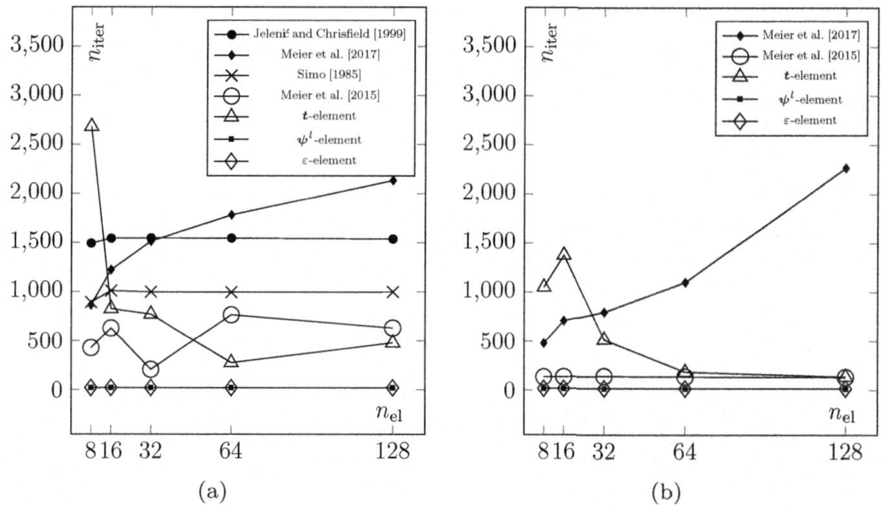

Fig. 3.14 Numerical example to study performance: The loading case illustrated in Fig. 3.13 is used to obtain the number of required Newton iterations in dependence of the number of elements for the new element types and different beam element formulations from the literature. The slenderness ratios. **a** $\rho = 100$ and **b** $\rho = 10{,}000$ are employed

[7] Figure reprinted with permission from Schulz & Böl (2019). ©2019 by Springer Nature

3.6.4.2 Cantilever Loaded by a Moment and a Force at Its End

To get a clearer picture of the performance characteristics another numerical example is studied. In more detail, the example shown in Fig. 3.6 is re-investigated. In this example, not only an external moment but also an external force loads the beam. This is an important note since the convergence order was found to depend on the type of loads and therefore the same might be true for the performance. The load adaption scheme is employed to determine the required iteration steps in dependence of the element number. The results are shown in Fig. 3.15.

The number of iterations of the $\boldsymbol{\psi}^l$-element and ε-element are independent of the number of elements and are between 35 and 50. That means that the performance is slightly worse when not only external moments are present as was the case with the load case depicted in Fig. 3.13. However, despite this small reduction in performance, these new element formulations are the most efficient ones among the tested beam formulations.

For the the $\boldsymbol{t}$-element no new findings are generated in this example. A higher element number is required before the performance approaches the one of the other beams. It seems noticeable that for high slenderness the simulation using 8 $\boldsymbol{t}$-elements did not converge.

The beam element formulations provided by Jelenić & Crisfield (1999) and Simo (1985) showed $n_{\text{iter}} \approx 7{,}500$ and $n_{\text{iter}} \approx 6{,}400$, respectively, for high slenderness and are not shown in Fig. 3.15b.

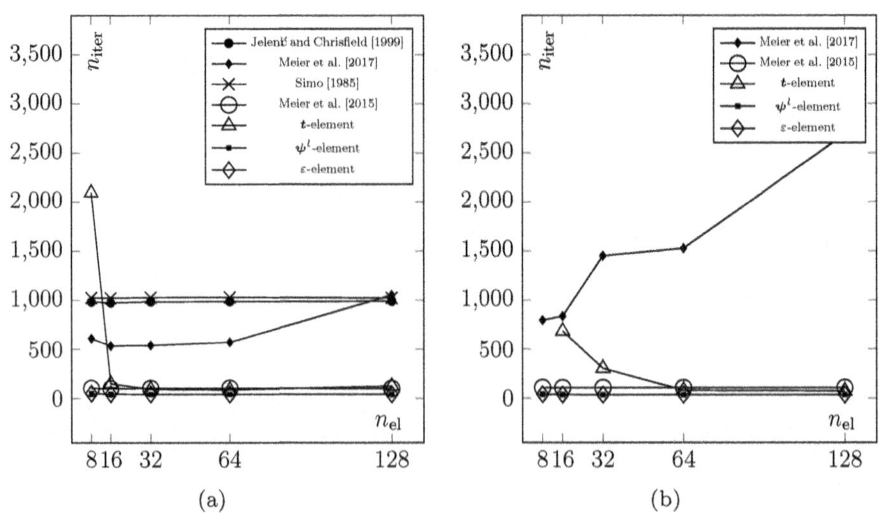

Fig. 3.15 Numerical example to study performance: For the loading case illustrated in Fig. 3.6 the number of required Newton iterations is plotted against the number of elements for various beam element formulations from the literature. Two different slenderness ratios. **a** $\rho = 100$ and **b** $\rho = 10{,}000$ are employed

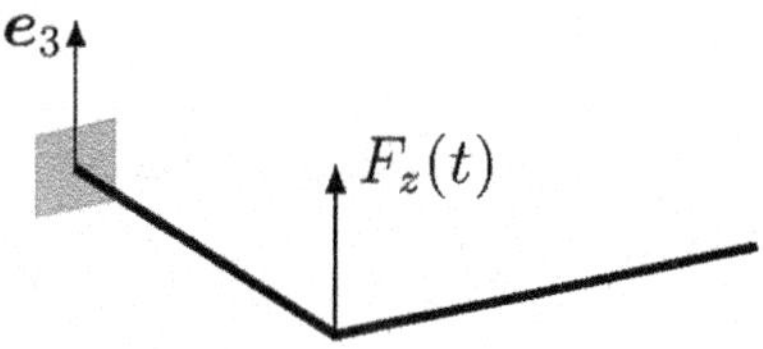

Fig. 3.16 Numerical dynamic example to test conservation of system energy of the ε-element: A right-angle elbow cantilever is loaded with a time-dependent force F_z at its connection

It is stressed that the superior Newton-Raphson performance of the ψ^l-element and ε-element is not due to favorable deformed configurations which exhibit mainly constant curvature. To further disperse this concern, the example shown in Fig. 3.10 was re-visited. It was found that, using 16 ψ^l-elements, a single loading step and a total number of 5 N-Raphson iterations were required, which is distinctly less than found for other beam formulations (Meier et al., 2017).

3.6.5 *Free Oscillation*

This final dynamic test case is commonly studied in the literature of geometrically nonlinear beam elements (Meier et al., 2017). Only the ε-element is considered since it seems to be the most promising beam formulation. In detail, two initially straight beams of length L_0 with $\rho = 100$ are rigidly connected at their ends to form a right-

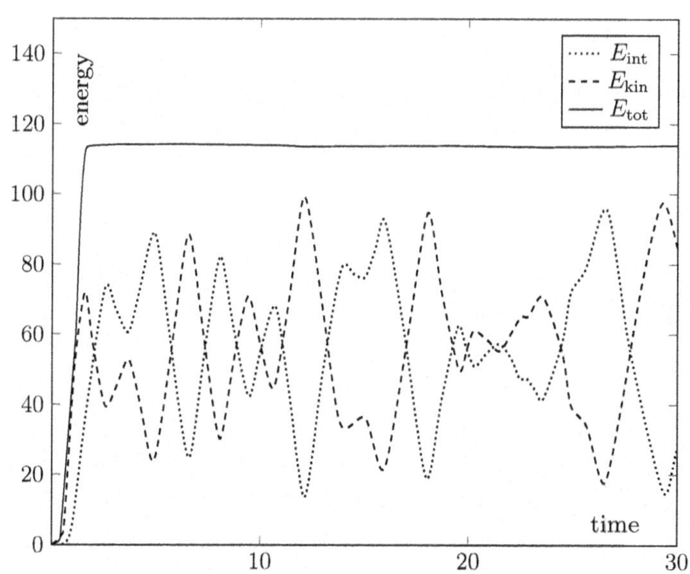

Fig. 3.17 Numerical dynamic example to test conservation of system energy of the ε-element: The kinetic, internal, and system energy are plotted over time. 2 ε-elements and $\Delta t = 0.25$ are employed

angle elbow cantilever of length $2L_0$. One end of this structure is clamped. The load $\overline{f}(s_0 = L_0, t) = (0, 0, F_z(t))^T_{e_i}$ acts at the connection and varies over time t. The load is linearly increased from $F_z = 0$ at $t = 0$ to $F_z = 50$ at $t = 1$ and subsequently linearly decreased to $F_z = 0$ at $t = 0$, see Fig. 3.16.

The resulting kinetic and internal energies E_{kin} and E_{int} as well as the total system energy E_{tot} are plotted over time until $t_{\text{end}} = 50$. A rough time step size and a sparse spatial discretizations are employed. $\Delta t = 0.25$ is paired with 2 ε-elements. The results are shown in Fig. 3.17.

It can be observed that the system energy is well preserved even for a rough discretization and for relatively large time steps. This might be attributed to the Lie group extension of the general Newmark time integration scheme used. As a big disadvantage of the implicit definition of the center-line, the global stiffness matrix is not sparse as usually is the case in finite element formulations. More thorough studies of the dynamic behavior of the new beam formulations should be conducted in future works.

Chapter 4
Modeling the Mechanics of Single Polymer Chains in the Finite-Element Framework

The Kirchhoff-Love beam and its finite element formulation presented in Sects. 2 and 3 are employed to model the mechanical behavior of polymers. In the remainder of this section, a model for a single polymer chain is presented including both contributions, intrinsic stiffness and temperature effects. A solution procedure within the finite-element framework is devised and used to fit the model to experimental data using a surrogate model based on a Gaussian process. Obtained material parameters are compared to the literature.

4.1 Basic Model Assumptions

The polymers considered here consist of covalently bonded monomers. They form a long and unbranched molecule dissolved in a Newtonian fluid. In many synthetic and biological materials such polymers constitute a main component (Mark, 2007; Pittman & Culbertson, 2012). Often, the flexibility or rigidity of a single chain is its most important physical property and determines its technological or biological usability (Rubinstein & Colby, 2003). Therefore, polymers are commonly categorized based on that property. This categorization is briefly introduced here.

Due to thermal effects the polymer chain is subject to collisions with molecules of the surrounding fluid forcing the polymer into a deformed state. Opposed to that, the intrinsic stiffness and the potentially non-straight initial configuration of the polymer influences the deformed configuration resulting from these collisions as well. If none of these two effects can be neglected the polymer is said to be semiflexible (Fixman & Kovac, 1973; Saitô et al., 1967). The category of semiflexible polymers contains many biologically relevant filaments such as DNA, actin, microtubules, and intermediate filaments. In the extreme case where the stiffness of the chain can be disregarded the polymer is said to be flexible and it might be modeled by so-called Gaussian chains

M. C. Schulz, *A New Kirchhoff-Love Beam Element and its Application to Polymer Mechanics*, Mechanics and Adaptronics,
https://doi.org/10.1007/978-3-031-06340-4_4

(Flory, 1989). A distinct experimental finding for polymers of this category is that a nonlinear mechanical response is found at high strains, only (Yamakawa, 1971). On the other end of the spectrum, where thermal effects are neglected, rigid polymers can be found. Sometimes, non-rigid polymers under high external stress are treated as rigid (e.g. Holzapfel & Ogden, 2011; MacKintosh, 2006).

Shear deformation is often neglected when modeling polymer mechanics. This is mainly justified by the high slenderness of polymer chains. Furthermore, a noticeable impact on physical quantities was not found in careful simulations. On top of that, there is no experimental data in this regard. Inertia forces are disregarded as well since they are exceeded by the friction forces caused by collisions with the surrounding fluid (Cyron & Wall, 2012).

When polymers exceed a critical length, i.e. when it surpasses a critical number of monomers $M = L_p^3/d^3$, self-contact or so-called self-avoidance becomes relevant (Strick et al., 2002). Applied to a single DNA molecule with a diameter $d = 2\,\text{nm}$ and a persistence length $L_p = 50$ nm, this means that $M = 15{,}625$. But self-contact is not considered here, since the simulated molecules are shorter than the critical length. But including this effect could be an interesting field of future research.

So-called bead-spring or bead-rod models (Alder & Wainwright, 1959; Rahman, 1964) have been used to study polymers in general as well as for the special case of single chains under external loads (e.g. Alexander-Katz et al., 2006; Mai et al., 2018). Comprehensive reviews (Gooneie et al., 2017; Karatrantos et al., 2016; Park et al., 2016) were dedicated to these Brownian dynamics simulations and their different implementations. Despite this long tradition, Cyron & Wall (2009) and Poelert et al. (2011) pointed out that this simulation approach has several drawbacks:

- absence of a mathematical basis
- requirement to add artificial constraints
- use of expensive explicit time integration
- difficulty to include other physical effects such as anistropic behavior or shear deformation

To address these shortcomings, Cyron & Wall (2009) proposed solving the governing stochastic partial differential equation (SPDE) within the finite-element framework using beam elements. The collisions with the molecules of the surrounding fluid were modeled as distributed loads, which are piece-wise constant in time and space. This approach was validated for semiflexible polymers by comparison with theory and experiments. For example, the tangent correlation function and the radial distribution function of an actin filament and of DNA could be reproduced. Using this method, which can be called a finite-element Brownian dynamics simulation, Poelert et al. (2011) studied shear-deformable and transverse isotropic microtubules and Cyron et al. (2013), Müller et al. (2015) and Kachan et al. (2016) even simulated polymer networks consisting of thousands of polymer chains. In a related work, Lin et al. (2014) presented a framework for small deformations.

In the present work, the governing SPDE is solved using the ε-element exploiting the fact that shear deformation can be neglected. In addition, the big disadvantage of that element, a full stiffness matrix in the dynamic case, does not matter since inertia forces can be disregarded. Furthermore, simulations corresponding to experimentally feasible constraints are performed in a Monte Carlo manner and the results are used to build a surrogate model to speed up the fitting process. As surrogate model a non-parametric regression method, a so-called Gaussian process, is employed. It learns the relationship between an input vector, the model parameters describing the mechanical behavior of the polymer chain, and the output, the displacement. With this fast procedure at hand experimental data can be fitted and even correlations among the parameters can be identified. The results of the parameter identification are compared with the literature.

4.2 Probabilistic Basics

In this section, the common basics of probabilistic notation are introduced as may be found in e.g. Bishop (2006) and Barber (2012).

Univariate Gaussian or normal distribution: Let y denote an observation of a real-valued random variable, which is called normally distributed if its probability density function is defined by

$$p(y|\mu, \sigma) = \frac{1}{\sigma\sqrt{2\pi}} \exp\left[-\frac{(y-\mu)^2}{2\sigma^2}\right], \tag{4.1}$$

where

$$\mu(y) = \mathbb{E}\left[p(y)\right] = \int y p(y) \mathrm{d}y \tag{4.2}$$

is the expected value (first moment of the distribution) and

$$\sigma^2(y) = \mathbb{V}\left[p(y)\right] = \mathbb{E}\left[(y - \mu(y))^2\right] \tag{4.3}$$

is the variance (second moment of the distribution). As a short hand

$$y \sim \mathcal{N}(\mu, \sigma^2) \quad \text{or} \quad y \sim \mathcal{N}(y|\mu, \sigma^2) \tag{4.4}$$

are introduced.

Multivariate or Joint Gaussian Distribution: The univariate case can be generalized to multivariate or joint distributions, where the probability density of a collection of observations of random variables $\boldsymbol{y} = [y_1, ..., y_N]$ then reads

$$p(\boldsymbol{y}|\boldsymbol{m}, \boldsymbol{K}) = \frac{1}{\sqrt{|\boldsymbol{K}|2\pi}} \exp\left[-\frac{1}{2}(\boldsymbol{y}-\boldsymbol{m})^T \boldsymbol{K}^{-1}(\boldsymbol{y}-\boldsymbol{m})\right], \tag{4.5}$$

if and only if the covariance matrix $\boldsymbol{K}$ is positive definite. Therein,

$$\boldsymbol{m}(\boldsymbol{y}) = (\mathbb{E}\left[p(y_1)\right], ..., \mathbb{E}\left[p(y_n)\right])^T \tag{4.6}$$

denotes the mean vector and $\boldsymbol{K}$ the covariance matrix with the elements

$$(\boldsymbol{K}(\boldsymbol{y}))_{ij} = \mathbb{E}\left[(p(y_i) - m_i)(p(y_j) - m_j)\right]. \tag{4.7}$$

As a short hand

$$\boldsymbol{y} \sim \mathcal{N}(\boldsymbol{m}, \boldsymbol{K}) \quad \text{or} \quad \boldsymbol{y} \sim \mathcal{N}(\boldsymbol{y}|\boldsymbol{m}, \boldsymbol{K}) \tag{4.8}$$

are written.

Marginal probability: If $\boldsymbol{y}_1$ and $\boldsymbol{y}_2$ denote two different collections of observations, the marginal probability of $\boldsymbol{y}_1$ is defined as

$$p(\boldsymbol{y}_1) = \int p(\boldsymbol{y}_1, \boldsymbol{y}_2)\mathrm{d}\boldsymbol{y}_2. \tag{4.9}$$

It is noted that if the joint distribution $p(\boldsymbol{y}_1, \boldsymbol{y}_2)$ is equal to the product of the marginals $p(\boldsymbol{y}_1) \cdot p(\boldsymbol{y}_2)$ those variables are independent.

Conditional probability: The probability of observing $\boldsymbol{y}_1$ given $\boldsymbol{y}_2$ is written as

$$p(\boldsymbol{y}_1|\boldsymbol{y}_2) = \frac{p(\boldsymbol{y}_1, \boldsymbol{y}_2)}{p(\boldsymbol{y}_2)}. \tag{4.10}$$

If $\boldsymbol{y}_1$ and $\boldsymbol{y}_2$ are independent, the marginal $p(\boldsymbol{y}_1)$ and the conditional $p(\boldsymbol{y}_1|\boldsymbol{y}_2)$ are equal.

Bayes' Theorem: Without proof it is given that

$$p(\boldsymbol{y}_1|\boldsymbol{y}_2) = \frac{p(\boldsymbol{y}_1)p(\boldsymbol{y}_2|\boldsymbol{y}_1)}{p(\boldsymbol{y}_2)}, \quad \text{where} \quad \text{posterior} = \frac{\text{prior} \cdot \text{likelihood}}{\text{marginal likelihood}}. \tag{4.11}$$

If the prior and the likelihood are Gaussian, the posterior is Gaussian as well and an analytical solution is available.

4.3 Viscous and Temperature Effects

In this section it is portrayed how to model the forces a polymer dissolved in a Newtonian fluid is subject to. The occurrence of collisions of the polymer with the molecules of the fluid may be modeled by viscous and random forces and moments

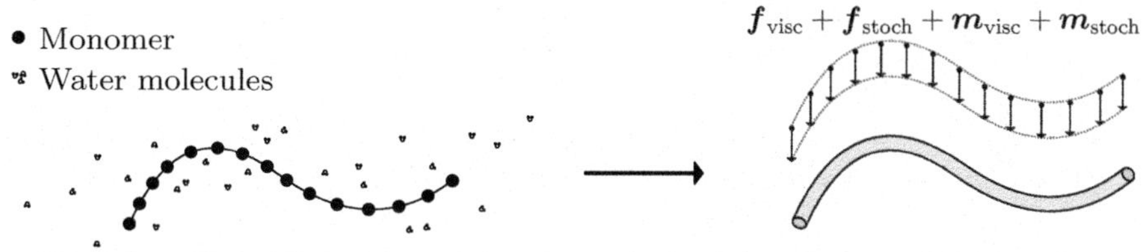

Fig. 4.1 On the *left side* a schematic representation of the molecular structure of a polymer dissolved in water is depicted. It is modeled as a beam under a distributed load as shown on the *right hand side* (adapted from Cyron & Wall (2012))

(Coffey et al., 2004). Cyron & Wall (2009) employed distributed loads to approximate these forces and moments, which are detailed in this section and sketched in Fig. 4.1.

4.3.1 Viscous Forces and Moments

The approach chosen here uses the standard notation of an anisotropic damping tensor. It distinguishes between perpendicular and parallel movement with regard to the center-line. Then, assuming a resting fluid, the viscous effects are represented as distributed loads and moments

$$\boldsymbol{f}_{\text{visc}} = \boldsymbol{\Lambda} \boldsymbol{c}_t \boldsymbol{\Lambda}^T \dot{\boldsymbol{r}} \quad \text{and} \quad \boldsymbol{m}_{\text{visc}} = \boldsymbol{\Lambda} \boldsymbol{c}_r \boldsymbol{\Lambda}^T \boldsymbol{W}, \tag{4.12}$$

respectively, where

$$\boldsymbol{c}_t = \begin{pmatrix} \gamma_{\parallel} & 0 & 0 \\ 0 & \gamma_{\perp} & 0 \\ 0 & 0 & \gamma_{\perp} \end{pmatrix}_{\boldsymbol{g}_i} \quad \text{and} \quad \boldsymbol{c}_r = \begin{pmatrix} \gamma_{\text{tor}} & 0 & 0 \\ 0 & 0 & 0 \\ 0 & 0 & 0 \end{pmatrix}_{\boldsymbol{g}_i} . \tag{4.13}$$

There are different friction models to determine $\gamma_{\parallel}$, $\gamma_{\perp}$, and γ_{tor} that have been applied to polymers in solution, such as models employing a single friction coefficient (Doi & Edwards, 1988), simplified formula when only limited data on the polymer are available (e.g. Doi & Edwards, 1988; Hallatschek et al., 2007), anisotropic approaches (Montesi et al., 2005), considerations based on the Navier-Stokes equation (Ermak & McCammon, 1978; Rotne & Prager, 1969), and models including long-range hydrodynamic interactions (Chandran & Mofrad, 2010). Also, cylindrical rods in laminar flows have been studied resulting in different friction coefficients for polymers (Bonet Avalos et al., 1993; Broersma, 1960; Ortega & Garcıa de la Torre, 2003). Here, the approach by Cyron & Wall (2012) is followed, which was shown to agree well with experiments and theory on semiflexible polymers. In this approach the friction coefficients introduced in Eq. (4.13) are inferred from diffusion coefficients as Ortega & Garcıa de la Torre (2003)

$$\gamma_{\parallel} = \frac{k_B T}{3LD_t - L^3 D_r/6}, \quad \gamma_{\perp} = \frac{12 k_B T}{L^3 D_r}, \quad \text{and} \quad \gamma_{\text{tor}} = \frac{k_B T}{L D_{\text{tor}}}, \tag{4.14}$$

where D_t depicts the translation diffusion coefficient, D_r is a coefficient related to rotation about an axis orthogonal to the center-line, and D_{tor} defines the translation diffusion coefficient. These three coefficients might be obtained from experiments, simulations, or analytical expressions. As presented by Cyron & Wall (2012), here, the latter method is chosen using analytical relations given in Ortega & Garcıa de la Torre (2003). It is noted that the damping matrices are evaluated at each Gauss point within the finite-element framework resulting in more accurate models compared to lumped solution procedures (Cyron & Wall, 2010).

4.3.2 Stochastic Forces and Moments

The approach chosen here is based on the assumption that the stochastic forces are a random Gaussian variable since they result from a large number of interactions. Thus, they are uniquely defined by their first and second moment, i.e. their mean and variance (Hallatschek et al., 2007)

$$\mathbb{E}[\boldsymbol{f}_{stoch}(s,t)] = 0 \tag{4.15}$$

$$\mathbb{V}[\boldsymbol{f}_{stoch}(s,t)] = 2k_B T \boldsymbol{\Lambda} \boldsymbol{c}_t \boldsymbol{\Lambda}^T \delta_{tt'} \delta_{ss'}, \tag{4.16}$$

respectively, where δ denotes the standard Dirac delta function. It is pointed out that the second moment is the scaled damping tensor. It follows that

$$\boldsymbol{f}_{stoch}(s,t) = \sqrt{2k_B T} \boldsymbol{\Lambda} \sqrt{\boldsymbol{c}_t} \frac{\partial^2 \mathcal{W}(s,t)}{\partial s \partial t}, \tag{4.17}$$

where $\mathcal{W}$ depicts three Brownian sheets packed into a vector. In contrast to a standard Wiener process, which depends only on time, the Brownian sheet depends on both, time and space.

It remains to reformulate the partial differentiation of the Brownian sheets to fit it into a finite-element framework. To this end, Allen et al. (1998) proposed a piecewise constant approximation

$$\begin{aligned} \frac{\partial^2 \hat{\mathcal{W}}}{\partial s \partial t} &= \sum_i \sum_j \frac{\chi_i(s)\chi_j(t)}{\Delta s \Delta t} \int_{t_i}^{t_{i+1}} \int_{s_i}^{s_{i+1}} \mathrm{d}\mathcal{W}(s,t) \\ &= \sum_i \sum_j \frac{\chi_i(s)\chi_j(t)}{\Delta s \Delta t} \sqrt{\Delta s \Delta t}\, \mathcal{N}(\boldsymbol{0}, \boldsymbol{1}), \end{aligned} \tag{4.18}$$

where $\mathcal{N}(\mathbf{0}, \mathbf{1})$ depicts a random three-dimensional Gaussian variable with zero mean and unit variance for all three random variables. Furthermore, to allow for concise notation

$$\chi_i(x) = \begin{cases} 1, & \text{if } x \in [i\Delta x; (i+1)\Delta x[\\ 0, & \text{otherwise} \end{cases} \tag{4.19}$$

is introduced. The stochastic integrals shown in Eq. (4.18) should be interpreted as backward or right-point integrals. That way, at constant temperature, noise-induced drift terms are avoided (Tian & Burrage, 2001). To further the theoretical foundation, the works of Allen et al. (1998) and Zhang & Karniadakis (2017) are cited who showed that the presented approximation converges to the right solution for linear and semi-nonlinear Dirichlet problems. In addition, Buckdahn & Pardoux (1990) proved the equivalence of the strong and weak form for SPDEs.

Combining these findings, the stochastic forces can be included in a finite element framework by adding

$$F_{stoch}[\varepsilon, \mathbf{\Lambda}](T) = \int_0^T \int_0^{L_0} \delta \boldsymbol{r}^T \boldsymbol{f}_{stoch} \mathrm{d}s_0 \approx \sum_i \chi_i(t) \sqrt{\frac{2k_B T}{\Delta s \Delta t}} \int_0^{L_0} \sum_j \chi_j(s) \delta \boldsymbol{r}^T \mathbf{\Lambda} \sqrt{\boldsymbol{c}_t} \boldsymbol{N}_{ij}(0, 1) \mathrm{d}s_0 \tag{4.20}$$

to the weak form defined in Eq. (2.128). The index in $\boldsymbol{N}_{ij}$ means that a different random draw from a Gaussian distribution is performed at every time and space increment. Naturally, that means that for every draw a different force is generated and thus a different deformed configuration is computed. That means this method does not result in mean values but single realizations according to certain random draws. The approximation is schematically depicted in Fig. 4.2.

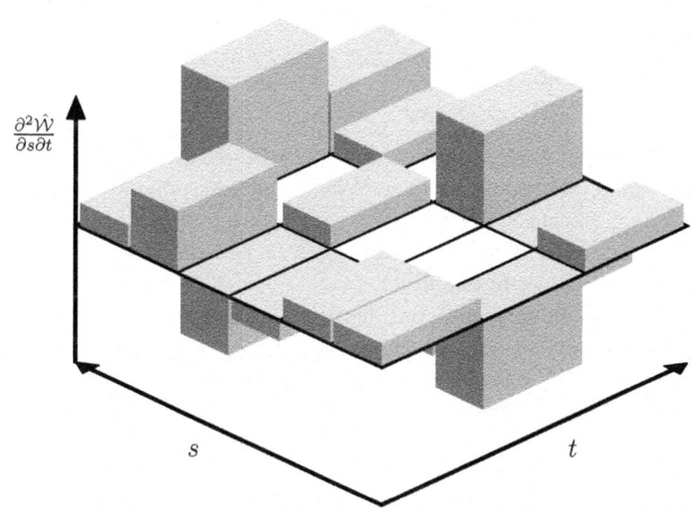

Fig. 4.2 Schematic depiction of the approximation of the Brownian sheet as spatially and temporally step-wise constant, normally distributed values

All arguments presented so far for stochastic forces can be analogously applied to stochastic moments resulting in

$$M_{stoch}[\varepsilon, \mathbf{\Lambda}](T) \approx \sum_i \chi_i(t)\sqrt{\frac{2k_BT}{\Delta s \Delta t}} \int_0^L \sum_j \chi_j(s)\delta \boldsymbol{r}^T \mathbf{\Lambda}\sqrt{c_r}\mathcal{N}_{ij}(\mathbf{0}, \mathbf{1})ds, \quad (4.21)$$

which needs to be added to the weak form in Eq. (2.128). Finally, it is mentioned, that there are other approximation methods for the white noise, for instance the orthogonal series expansion (Zhang & Karniadakis, 2017). Further, Cyron & Wall (2009) found that the discrete damping matrix is equivalent to the covariance matrix of the discrete stochastic forces when the Bubnov-Galerkin method is employed. If it is not used a fine discretization s required to obtain a good approximate solution.

4.4 Constitutive Parameters and Discretization

Besides including the viscous and temperature effects defined in the last section, other quantities of the model still remain to be specified. In the following, the constitutive parameters and the spatial and temporal discretization are detailed.

Constitutive parameters: The constitutive parameters required in the beam element formulation, see Eq. (2.133), are given as

$$\boldsymbol{C}_M = \text{diag}\,[GI_t, EI_2, EI_3]_{e_i} \quad \text{for } i = 1, 2, 3, \quad (4.22)$$

which can be reformulated for polymers assuming an isotropic polymer chain (Le Goff et al., 2002)

$$EI_2 = EI_3 = k_BTL_p. \quad (4.23)$$

Herein, L_p denotes the persistence length, k_B defines the Boltzmann constant and T is the absolute temperature. A simple linear constitutive model is used assuming that locally only small deformations occur. When modeling polymers this is a common approach, see e.g. Shi et al. (1995). It allows for good agreement between simulations, theoretical results, and experiments. Nevertheless, non-linear material models are an interesting task for future research.

Also, as found in simulations, consideration of torsional deformation did not alter the probability density of the end-to-end distance or the force-extension behavior. Thus, torsional rigidity cannot be inferred from the following numerical methods and is computed from the known diameter of DNA $d = 2 \cdot 10^{-9}$ m assuming a solid circular cross-section.

Temporal discretization: The problem of modeling a polymer in a Newtonian fluid was achieved using a stiff right-point SPDE, which is expected to be stiff due to the high difference in bending and axial stiffness found for polymers. For such stiff SPDEs the Euler backward discretization method is not only beneficial but also easy to implement (Ermak & McCammon, 1978; Montesi et al., 2005; Tian & Burrage, 2001). Therefore, this solution scheme is employed and introduced here and not the one portrayed in Sect. 3.1.3.

Considering the current time step t^n and the last time step t^{n-1} the temporal derivatives are approximated by

$$\dot{\varepsilon}^n \approx \frac{\varepsilon^n - \varepsilon^{n-1}}{\Delta t} \quad \text{and} \quad \dot{\boldsymbol{\Lambda}}^n = \boldsymbol{\Lambda}^n \boldsymbol{S}(\boldsymbol{W}^n) \approx \boldsymbol{\Lambda}^n \boldsymbol{S}\left(\frac{\Delta\tilde{\boldsymbol{\Theta}}}{\Delta t}\right). \tag{4.24}$$

It is repeated here that $\Delta\tilde{\boldsymbol{\Theta}}$ and $\Delta\tilde{\boldsymbol{\theta}}$ are the material and spatial rotation vector increment between two time points, respectively. Risking slight repetition, the update strategy at time t^n

$$\Delta\boldsymbol{\Lambda}^{k+1,n} = \Delta\boldsymbol{\theta}\boldsymbol{\Lambda}^{k,n} \quad \text{with} \quad \Delta\tilde{\boldsymbol{\theta}} = \boldsymbol{T}(\tilde{\boldsymbol{\theta}})\Delta\boldsymbol{\theta}, \tag{4.25}$$

is re-formulated, where (Jelenić & Crisfield, 1999)

$$\boldsymbol{T}(\tilde{\boldsymbol{\theta}}) = \frac{1}{\tilde{\theta}}\tilde{\boldsymbol{\theta}} \otimes \tilde{\boldsymbol{\theta}} + \frac{\tilde{\theta}/2}{\tan(\tilde{\theta}/2)}(\boldsymbol{I} - \frac{1}{\tilde{\theta}^2}\tilde{\boldsymbol{\theta}} \otimes \tilde{\boldsymbol{\theta}}) - \frac{1}{2}\boldsymbol{S}(\tilde{\boldsymbol{\theta}}) \tag{4.26}$$

relates an increment between two different time steps and an increments between different configuration at the same time step.

Spatial discretization: To solve the spatial integral in Eq. (4.20) using the standard Newton-Cotes approach, Cyron & Wall (2010), Cyron & Wall (2012), and Poelert et al. (2011) chose a non-constant Δs that varies in such a way that the same standard Gauss-Legendre integration could be applied that was used to integrate over the local coordinate ξ. In a different approach, Lin et al. (2014) chose to set the element length equal to Δs. Both approaches fail for more flexible polymers since large deformation gradients arise across the spatial intervals, each of which exhibits a different value for the line load. To address this problem here Δs is bigger than the element length requiring slightly more effort to realize the integral over the stochastic forces and moments.

4.5 Surrogate Modeling

Ultimately, the goal is to fit the present model to experimental data. But the model presented so far needs to be used in a Monte Carlo manner to obtain mean values of quantities of interest. That means that to obtain the extension of a single curve under external load, the simulation must be continued until a mean value with acceptable variance is reached. To vastly accelerate the fitting process, a surrogate model will be constructed that directly computes the mean values for given material parameters.

There are different classes of surrogate models. They can roughly be grouped in data-fit methods, hierarchical methods, and reduced order models (Eldred et al., 2004; Robinson et al., 2008). The latter two approaches aim at capturing underlying physical effects, while the first method solely tries to perform a regression on available data. Such a regression can be performed using different strategies, such as neural networks, polynomial chaos expansion, support vector machines, and Gaussian processes (Asher et al., 2015). Liu et al. (2014) used a Gaussian process to reproduce the results of a time-consuming model. The surrogate model was then used to find parameters of the original model by data fitting. Based on that example in this work, too, a Gaussian process is applied. The corresponding basic theory is described in the following based on Barber (2012), Bishop (2006), and Rasmussen & Williams (2006). For better understanding it is started by introducing linear regression models, which can be expanded to Gaussian processes.

4.5.1 Linear Model for Regression

Let $(\boldsymbol{x}_n, y_n)$ be pairs of observations, where $\boldsymbol{x}$ is the independent input vector and y the scalar output. These quantities can be assembled to construct $\boldsymbol{X} = [\boldsymbol{x}_1, ..., \boldsymbol{x}_N]$ and $\boldsymbol{y} = [y_1, ..., y_N]$ with N being the total number of observations. $D = \{\boldsymbol{X}, \boldsymbol{y}\}$ denotes the whole data. In the scope of this work, performing a regression or a fit then refers to inferring a mapping from $\boldsymbol{x}$ to y based on observed data. A large class of viable functions for linear regression with scalar output can be summarized as

$$f(\boldsymbol{x}) = \boldsymbol{w}^T \boldsymbol{\phi}(\boldsymbol{x}), \tag{4.27}$$

where $\boldsymbol{w}$ is a vector containing the parameters of the model and $\boldsymbol{\phi}(\boldsymbol{x})$ is a *basis* or *feature function*. In the easy case of a straight line fit for scalar-valued input x and output it holds that

$$\boldsymbol{\phi}(x) = (1, x)^T \quad \text{and} \quad \boldsymbol{w} = (a, m)^T, \tag{4.28}$$

where a and m denote intercept and slope, respectively.

If y is considered to be a random variable, the linear model is extended by a noise term ε to obtain

$$y = f(\boldsymbol{x}) + \varepsilon. \tag{4.29}$$

Commonly $\varepsilon \sim \mathcal{N}(0, \sigma_\varepsilon^2)$ is assumed but other distributions might be sensible depending on the underlying physical problem.

The method of least squares and regularization: Disregarding the noise term for the moment, the parameters in $\boldsymbol{w}$ can be obtained by minimizing the sum of squared differences

$$E(\boldsymbol{w}) = \sum_{n=1}^{N} (y_n - \boldsymbol{w}^T \boldsymbol{\phi}(\boldsymbol{x}_n))^2 = |\boldsymbol{y} - \boldsymbol{w}^T \boldsymbol{\phi}(\boldsymbol{X})|^2, \tag{4.30}$$

where in a small abuse of notation the stacked feature vector

$$\boldsymbol{\phi}(\boldsymbol{X}) = [\boldsymbol{\phi}(\boldsymbol{x}_1), ..., \boldsymbol{\phi}(\boldsymbol{x}_N)] \tag{4.31}$$

is introduced. To prevent overly complex models leading to over-fitting and to make sure that the fitted model generalizes well a *regularizing* or *penalty term* is added. Typically this procedure results in the error function

$$E'(\boldsymbol{w}) = E(\boldsymbol{w}) + \lambda \boldsymbol{w}^T \boldsymbol{w} \tag{4.32}$$

and the additional problem of finding the regularizing parameter λ arises.

Maximizing the likelihood: Taking into consideration the Gaussian error term, the likelihood of finding the experimental observations $\boldsymbol{y}$ according to the model is

$$p(\boldsymbol{y}|\boldsymbol{X}, \boldsymbol{w}, \sigma_\varepsilon) = \prod_{n=1}^{N} \mathcal{N}(y_n|\boldsymbol{w}^T \boldsymbol{\phi}(\boldsymbol{x}_n), \sigma_\varepsilon^2). \tag{4.33}$$

Here, it is assumed that all data pairs are generated identically and independently. Furthermore, Gaussian distributions are closed under multiplication such that

$$p(\boldsymbol{y}|\boldsymbol{X}, \boldsymbol{w}, \sigma_\varepsilon) = \frac{1}{(2\pi\sigma_\varepsilon^2)^{\frac{N}{2}}} \exp\left[-\frac{1}{2\sigma_\varepsilon^2}|\boldsymbol{y} - \boldsymbol{w}^T \boldsymbol{\phi}(\boldsymbol{X})|^2\right] = \mathcal{N}(\boldsymbol{w}^T \boldsymbol{\phi}(\boldsymbol{X}), \sigma_\varepsilon^2 \boldsymbol{I}). \tag{4.34}$$

Typically, the parameters are identified by maximizing the logarithm of the likelihood

$$\ln p(\boldsymbol{y}|\boldsymbol{X}, \boldsymbol{w}, \sigma_\varepsilon) = -N \ln \sigma_\varepsilon - \frac{N}{2} ln(2\pi) - \sigma_\varepsilon^{-2} E(\boldsymbol{w}). \tag{4.35}$$

It is instructive to recognize that in the probabilistic framework the choice of additive Gaussian noise can be identified with the choice of a least squares error function. Furthermore, maximizing the likelihood gives the mode of the posterior. If the posterior is symmetric this coincides with the first moment of the distribution.

As before, a mere maximization of the likelihood (by using different basis functions) leads to complex models and over-fitting and regularization terms should be added.

Bayesian approach: The previous two methods give only point-estimates for the parameters. But due to often very limited data sets, these estimates are uncertain. In addition, the problem of overly complex models needs to be addressed artificially by adding a penalizing term. A Bayesian treatment tackles these problems by using Bayes' theorem given in Sect. 4.2. A common choice for a prior, when limited initial knowledge is available, is a multivariate zero-mean isotropic normal distribution

$$p(\boldsymbol{w}|\alpha) = \mathcal{N}(\boldsymbol{w}|\boldsymbol{0}, \alpha^{-1}\boldsymbol{I}). \tag{4.36}$$

In other cases, priors can be used to formulate the initial belief on the parameters based on

- physics,
- mathematical or physical models,
- expert's judgment,
- databases.

α and σ_ε are referred to as hyperparameters. The additive Gaussian noise leads to a Gaussian likelihood distribution $p(\boldsymbol{y}|\boldsymbol{X}, \boldsymbol{w}, \sigma_\varepsilon)$ as stated above. In this case, an analytical solution of Bayes' theorem for the posterior is available as

$$p(\boldsymbol{w}|D, \alpha, \sigma_\varepsilon) = \frac{p(\boldsymbol{y}|\boldsymbol{X}, \boldsymbol{w}, \sigma_\varepsilon)p(\boldsymbol{w}|\alpha)}{p(\boldsymbol{y}|\boldsymbol{X})} = \mathcal{N}(\boldsymbol{w}|\boldsymbol{m}_P, \boldsymbol{\Sigma}_P) \tag{4.37}$$

with

$$\boldsymbol{m}_P = \sigma_\varepsilon^{-2}\boldsymbol{\Sigma}_P\boldsymbol{\phi}(\boldsymbol{X})\boldsymbol{y} \quad \text{and} \quad \boldsymbol{\Sigma}_P^{-1} = \alpha\boldsymbol{I} + \sigma_\varepsilon^{-2}\boldsymbol{\phi}(\boldsymbol{X})\boldsymbol{\phi}(\boldsymbol{X})^T. \tag{4.38}$$

The logarithm of the posterior reads

$$\ln p(\boldsymbol{w}|D, \alpha, \sigma_\varepsilon) = -\frac{\sigma_\varepsilon^{-2}}{2}E(\boldsymbol{w}) - \frac{\alpha}{2}\boldsymbol{w}^T\boldsymbol{w} + \text{const.} \tag{4.39}$$

Thus, it can be seen that the prior naturally induces a regularizing term.

To obtain the predictive distribution at a new input $\boldsymbol{x}^*$ (what we are really interested in) the likelihood times the posterior needs to be marginalized over $\boldsymbol{w}$ resulting in

$$p(y^*|\boldsymbol{x}^*, D, \alpha, \sigma_\varepsilon) = \int p(\boldsymbol{y}|\boldsymbol{x}^*, \boldsymbol{w}, \sigma_\varepsilon)p(\boldsymbol{w}|D, \alpha, \sigma_\varepsilon)\mathrm{d}\boldsymbol{w} = \mathcal{N}(y^*|\boldsymbol{m}_P^T\boldsymbol{\phi}(\boldsymbol{x}^*), \sigma_{y^*}) \tag{4.40}$$

with

$$\sigma_{y^*} = \sigma_\varepsilon^2 + \boldsymbol{\phi}(\boldsymbol{x}^*)\boldsymbol{\Sigma}_P\boldsymbol{\phi}(\boldsymbol{x}^*)^T. \tag{4.41}$$

The remaining task is to determine α and σ_ε, which is done here using the evidence procedure. Other methods are available in the literature.

Evidence procedure: In a fully Bayesian treatment priors would be required not only over the parameters $\boldsymbol{w}$ but also over the hyperparameters α and σ_ε. But unfortunately a marginalization over all these quantities is generally not analytically traceable. Therefore, we turn to the evidence approximation

$$p(y^*|\boldsymbol{x}^*, D) \approx p(y^*|\boldsymbol{x}^*, D, \hat{\alpha}, \hat{\sigma_\varepsilon}). \tag{4.42}$$

That means we assume that the posterior $p(\alpha, \sigma_\varepsilon)$ is peaked around $\hat{\alpha}$ and $\hat{\sigma_\varepsilon}$ such that an integral over the hyperparameters is unnecessary and simply plugging in these values is accurate enough. Thus, the goal is to maximize the likelihood of observing the experimental findings according to the model marginalized over $\boldsymbol{w}$, which is

$$p(\boldsymbol{y}|\boldsymbol{X}, \hat{\alpha}, \hat{\sigma_\varepsilon}) = \int p(\boldsymbol{y}|\boldsymbol{X}, \boldsymbol{w}, \hat{\sigma_\varepsilon}) p(\boldsymbol{w}|\hat{\alpha}) \mathrm{d}\boldsymbol{w}. \tag{4.43}$$

This integral can be solved analytically. The logarithm of the solution reads

$$\ln p(\boldsymbol{y}|\boldsymbol{X}, \hat{\alpha}, \hat{\sigma_\varepsilon}) = \ln \hat{\alpha} - N \ln \sigma_\varepsilon - \frac{N}{2} ln(2\pi) - E'(\boldsymbol{m}_P) + \frac{\hat{\alpha}}{2} \boldsymbol{m}_P^T \boldsymbol{m}_P - \frac{1}{2} \ln \det \boldsymbol{\Sigma}_P^{-1}, \tag{4.44}$$

where

$$E'(\boldsymbol{m}_P) = \frac{\sigma_\varepsilon^{-2}}{2} ||\boldsymbol{y} - \boldsymbol{\phi}(\boldsymbol{X})^T \boldsymbol{m}_P||^2 + \frac{\hat{\alpha}}{2} \boldsymbol{m}_P^T \boldsymbol{m}_P \tag{4.45}$$

is singled out to stress the similarity to regularized least-squares method. Maximizing this equation yields values for $\hat{\alpha}$ and $\hat{\sigma_\varepsilon}$. More complex priors exhibiting a higher amount of hyperparameters are possible within this framework.

4.5.2 Gaussian Process Regression

Studying Eq. (4.37) it becomes evident that the feature vector $\boldsymbol{\phi}(\boldsymbol{X})$ occurs only in scalar products. Thus, the kernel trick can be applied. Instead of computing the feature vector and subsequently the scalar product, a kernel is used to directly compute the result of the scalar product. One way of deriving such a kernel for a Gaussian process is to increase the number of parameters to infinity and to compute the limit of the resulting scalar product. If that succeeds, no more parameters are required and the Gaussian process is therefore called non-parametric.

As it was also the case with linear models for regression presented in the last section, at every input vector $\boldsymbol{x}$ of dimension d a Gaussian distribution over the output is defined. Remembering that all finite combinations of Gaussians are still Gaussian the Gaussian process model can be summarized as

$$\boldsymbol{y}(\boldsymbol{X}) \sim \mathcal{N}(\boldsymbol{m}_X, \boldsymbol{K}(\boldsymbol{X}, \boldsymbol{X})). \tag{4.46}$$

Herein, $\boldsymbol{m}_X \in \mathbb{R}^N$ depicts the mean vector and $\boldsymbol{K}(\boldsymbol{X}, \boldsymbol{X}) \in \mathbb{R}^{N\times N}$ is the covariance matrix with

$$(\boldsymbol{K})_{ij} = \mathbb{E}\left[(p(y_i) - \boldsymbol{m}_i)(p(y_j) - \boldsymbol{m}_j)\right], \tag{4.47}$$

which describes the covariance for an output with itself and between different outputs. The inversion of the kernel matrix exhibits a complexity of $\mathcal{O}(N^3)$ and dominates the computation time.

It can be seen why a Gaussian process is defined to be a collection of random variables. It is the marginalization of an infinite-dimensional Gaussian distribution down to a finite number of random variables. The difference to the linear models described in the last section is that here, outputs are not described by any parameters but by their covariance with all other data and with itself. This can be better understood when looking at how to compute new output.

Mean and variation for new input: The distribution of a new output y^* for a new input vector $\boldsymbol{x}^*$ can be computed via

$$p(y^*|\boldsymbol{X}, \boldsymbol{y}, \boldsymbol{x}^*) = \int p(\boldsymbol{f}_*|\boldsymbol{X}, \boldsymbol{x}_*, \boldsymbol{f})p(\boldsymbol{f}|\boldsymbol{X}, \boldsymbol{y})\mathrm{d}\boldsymbol{f}, \tag{4.48}$$

which can be given analytically as

$$\mathbb{E}\left[p(y^*|\boldsymbol{X}, \boldsymbol{y}, \boldsymbol{x}^*)\right] = m_* + \boldsymbol{k}_*^T(\boldsymbol{K} + \sigma_n^2\boldsymbol{I})^{-1}(\boldsymbol{y} - \boldsymbol{m}) \tag{4.49}$$

and

$$\mathbb{V}\left[p(y^*|\boldsymbol{X}, \boldsymbol{y}, \boldsymbol{x}^*)\right] = k(\boldsymbol{\theta}, \boldsymbol{x}^*, \boldsymbol{x}^*) - \boldsymbol{k}_*^T(\boldsymbol{K} + \sigma_n^2\boldsymbol{I})^{-1}\boldsymbol{k}_*, \tag{4.50}$$

respectively. It can be seen why the mean value is called a linear predictor since it linearly combines the training data points. Thus, it remains to define a method to compute the covariance matrix.

Covariance functions: The covariance matrix is approximated by a covariance function k such that

$$(\boldsymbol{K})_{ij} \approx k(\boldsymbol{x}_i, \boldsymbol{x}_j). \tag{4.51}$$

The resulting $\boldsymbol{K}$ must be positive semi-definite. Covariance functions most often work on the difference between input functions $(\boldsymbol{x}_i - \boldsymbol{x}_j)$ or on their similarity $(\boldsymbol{x}_i \cdot \boldsymbol{x}_j)$. If those operations are not defined for the input vectors, they need to be transformed to another space where those operations are indeed defined. Details on how to construct kernels and how to analyze their properties, the reader is referred to Rasmussen & Williams (2006). See Duvenaud (2014) and references therein on how to choose a covariance function. In general, a covariance function is studied with regard to its variability and differentiability. A typical choice is the Matérn-(5/2) covariance function

$$k(\boldsymbol{x}, \boldsymbol{x}') = \sigma_{\mathrm{mat}}^2 \left[\prod_{j=1}^{d}\left(1 + \frac{\sqrt{5}h_j}{l_j} + \frac{5h_j^2}{3l_j^2}\right)\right] \exp\left(-\sqrt{5}\sum_{j=1}^{3}\left(\frac{h_j}{l_j}\right)\right), \tag{4.52}$$

where σ_{mat}^2 is signal variance, h_j defines the absolute difference between the jth element of $\boldsymbol{x}$, and l_j depicts the correlation lengths. It is pointed out that $k(\boldsymbol{x}, \boldsymbol{x}) = \sigma_{\mathrm{mat}}^2$. This covariance function was shown to adequately represent the training data, see Sect. 4.6.

Log marginal likelihood: Using the evidence procedure described in Sect. 4.5.1, the logarithm of the marginal likelihood

$$\log p(\boldsymbol{y}|\boldsymbol{X}) = -0.5(\boldsymbol{y}-\boldsymbol{m})^T(\boldsymbol{K}+\sigma_\varepsilon^2\boldsymbol{I})^{-1}(\boldsymbol{y}-\boldsymbol{m}) - 0.5\log\left[\det\left(\boldsymbol{K}+\sigma_\varepsilon^2\boldsymbol{I}\right)\right] - \frac{n}{2}\log(2\pi) \tag{4.53}$$

is optimized to find the hyperparameters. This approach constitutes a compromise between model complexity and extendibility to new data points.

Specification for the current application: The material parameters and the external force form the input vector

$$\boldsymbol{x} = [L_0, L_p, EA, \boldsymbol{f}_{\mathrm{ext}}]^T, \tag{4.54}$$

where the external force vector contains only a single non-zero value. The scalar output is the end-to-end distance R, see Fig. 4.3 for an illustration. Since the training data is generated from simulations it is assumed to be noise-free (Rasmussen & Williams, 2006), i.e. $\sigma_\varepsilon = 0$. The hyperparameter of interest are thus

$$\boldsymbol{\theta} = [\sigma_{\mathrm{mat}}, h_j, l_j]^T \quad \text{for} \quad i = 1, 2, 3, 4. \tag{4.55}$$

A conjugate-gradient optimizer is used in combination with a latin-hypercube sampling strategy to find the hyperparameters corresponding to the global maximum of the marginal log-likelihood. For further stabilization of the optimization process, the Laplace approximation is used as described in Rasmussen & Williams (2006). Finding more sophisticated covariance functions including prior knowledge of the problem might be an interesting field of research for future work.

4.6 Numerical Results

The finite-element Brownian dynamics simulation method presented so far is used to perform a fit to experimental data and to determine material parameters. First, it is studied how to obtain mean extension values of given accuracy for chains under external load. Secondly, using the results, a Gaussian process is trained to act as a surrogate model describing the force-extension behavior. Thirdly and lastly, a regression based on different experimental data is conducted in a Bayesian framework to identify L_0, L_p, and EA. The polymer is assumed to be submerged in water such that $T = 293.15$K and the viscosity is known to be $\eta = 0.001$Pa·s. In addition, the salt concentration is assumed to be high enough such that electrostatic effect can be neglected.

The discretization in time and space is chosen to be $\Delta t = 10^{-5}$ s and $\Delta s \approx 5 \cdot 10^{-8}$ m based on results by Cyron & Wall (2012). In cases of divergence the element length is halved. To allow the simulation to reach a representative configuration, an initial equilibration time equal to the slowest bending eigenmode (Poelert et al., 2011)

$$t_{min} = \frac{\gamma_{\perp}}{k_B T L_p} q^{-4}, \tag{4.56}$$

is employed. All data points from this first time period are disregarded. Therein, q is a parameter depending on the boundary conditions, where $q = \pi/L_0$ for free ends, $q = 1.5\pi/L_0$ for hinged ends, and $q = 1.875\pi/L_0$ for a beam fixed on one end and free on its other end (Wiggins et al., 1998).

Furthermore, a lag time between different sample points needs to be defined to prevent correlation. For this purpose, t_{min} can be used as well. Assuming $q = \pi/L_0$, $L_p = 50$ nm, and $L_0 = 680$ nm, which are common values of these quantities, would result in $t_{min} \approx 8 \cdot 10^{-5}$. Therefore, only every tenth time step is used towards the computation of mean values.

4.6.1 Distributions for Different Boundary Conditions and Loads

Depending on the experimental setup that is used to pull apart a polymer, different kinematic constraints or boundary conditions are present at its ends. An exemplary overview is schematically presented in Fig. 4.3. In the statistical mechanics theory of very long polymer chains, which is often used to interpret experimental data, the boundary conditions are commonly neglected (Marko & Siggia, 1995; Wang et al., 1997). But Salomo et al. (2006) noticed a reduction of Brownian motion when confining the DNA to the experimental apparatus. In addition, experiments are usually conducted with a high salt concentration such that electrostatic repulsion forces can be neglected. Purohit et al. (2008) theoretically deduced a dependence of the force-extension on the boundary conditions with restriction to small deformations.

To further the understanding in this regard and to study the finite element Brownian dynamics method, the experimental setup by Salomo et al. (2006) is simulated to obtain normalized probability densities over the end-to-end distance R. It is reiterated here that $\boldsymbol{f}_{stoch}$ and $\boldsymbol{m}_{stoch}$ are randomly generated at every time step, each value for it results in a single value for R upon convergence of the finite element analysis.

Salomo et al. (2006) immobilized one end of a single double-stranded DNA on a polystyrene bead, which is moved by so-called optical tweezers. They are optical traps able to exert forces $\boldsymbol{f}_{\text{ext}}$ but not moments. Therefore, they are represented as hinged ends in the simulation. The other end of the chain was bonded to a second

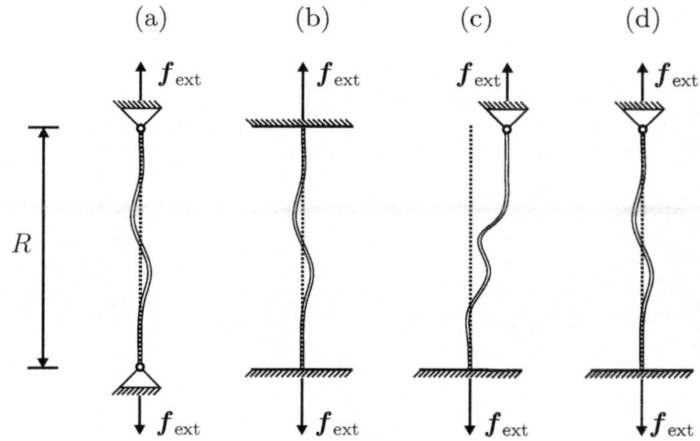

Fig. 4.3 Schematic representation of Dirichlet boundary conditions typically found in experiments performed to study polymer mechanics. The *triangles* represent fixed bearings and the *straight lines* symbolize clamps. Only in (**c**) lateral displacement is possible (adapted from Purohit et al. (2008))

polystyrene bead moved by a micropipette, which is able to exert forces and moments. It is modeled as fixed end.

Based on Wang et al. (1997), $L_0 = 680\,\text{nm}$ and $L_p = 52\,\text{nm}$ are used as approximations for material parameters of DNA and E can be estimated from the bending stiffness. For varying external forces and boundary conditions $N_{dp} = 100{,}000$ data points of the end-to-end distance R are generated. The results are used to determine probability density functions, as depicted in Fig. 4.4. In (a), no boundary conditions at all are present and the chain moves freely. In (b), the experimental setup used by Salomo et al. (2006) is recreated but without external force acting on the chain. In (c), the same conditions are present but $f_{\text{ext}} = 0.5\text{pN}$ is added. In (d), still the same conditions are maintained but the external force is increased to $f_{\text{ext}} = 1\text{pN}$.

The results clearly show that boundary conditions (and external forces, of course) influence the mean value of R. Therefore, it is important to consider them when interpreting force-extension data.

4.6.2 Tracking the Accuracy of Monte Carlo Methods

Ultimately, the mean value of the end-to-end distance $\bar{R}$ is of interest, not the result R_i at a certain time step. When determining $\bar{R}$ based on a large number of simulation results, different error sources can be identified: the Monte Carlo error MCE and the discretization of time, space, and the white noise. Due to the fine discretization the latter is negligible and not further studied.

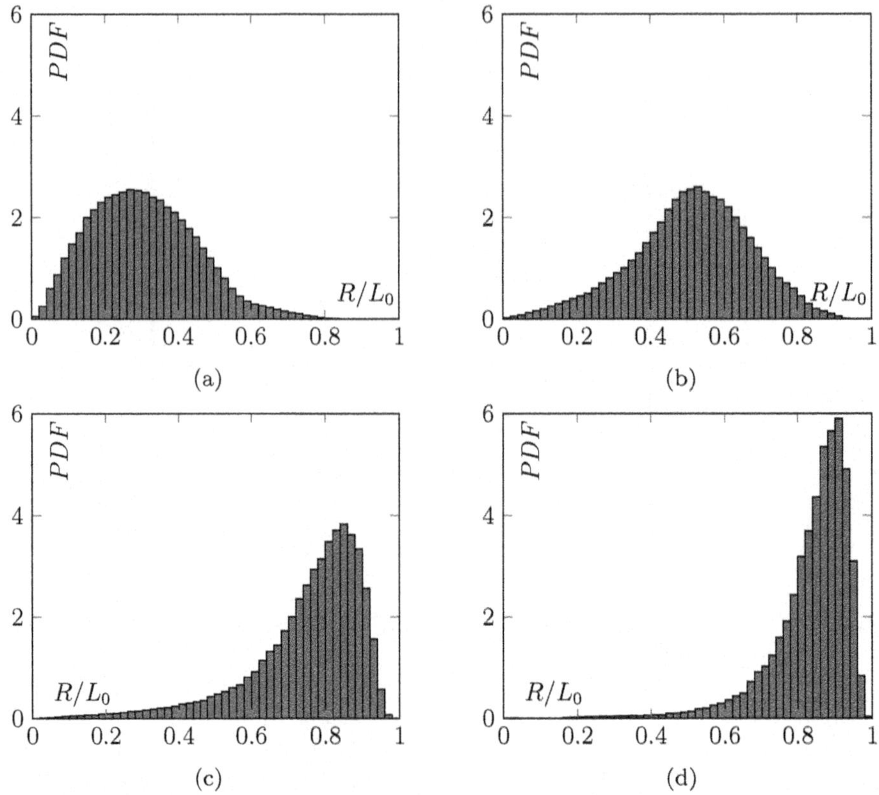

Fig. 4.4 Probability density functions over R/L_0 with (**a**) free ends, (**b**) constrained ends, (**c**) constrained ends and $f_{\text{ext}} = 0.5\text{pN}$, and (d) constrained ends and $f_{\text{ext}} = 1\text{pN}$[1]

Based on the approach presented in Koehler et al. (2009), two error measures are introduced. Firstly, the standard deviation of $\bar{R}$, which results from asymptotic theory (Robert & Casella, 2013)

$$MCE_{\text{mean}} = \frac{1}{N_{dp}} \sqrt{\sum_{i=1}^{N_{dp}} (R_i - \bar{R})^2} \tag{4.57}$$

can be regarded as MSE. This equation requires a good estimate of $\bar{R}$ a priori and is therefore not suited to track the accuracy of the MCE with increasing sample number. But it can be used to asses the accuracy of other methods.

Secondly, based on bootstrapping, a re-sampling method (Efron & Tibshirani, 1994)

[1] Figure reprinted with permission from Schulz et al. (2019). ©2019 by Elsevier

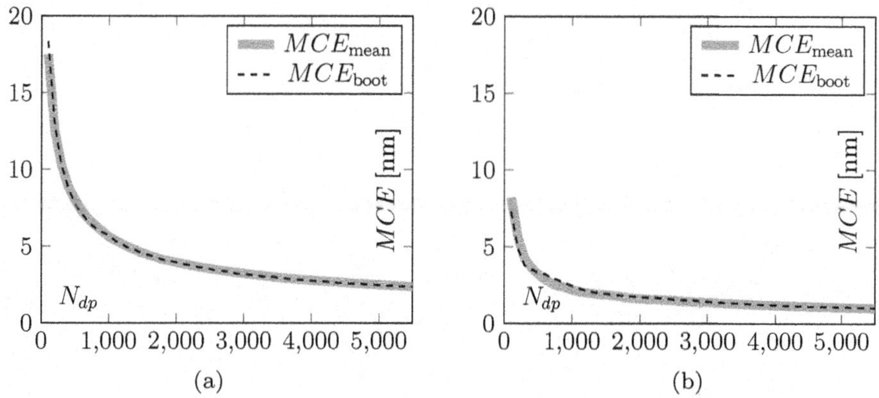

Fig. 4.5 MCE_{mean} and MCE_{boot} as defined in Eqs. (4.57) and (4.58), respectively, are plotted over data points N_{dp} in (**a**) and (**b**) drawn from the distributions shown in Fig. 4.4 (**a**) and (**d**), respectively[2]

$$MCE_{\text{boot}} = \sqrt{\frac{1}{B}\sum_{i=1}^{B}(\bar{R}_i - \bar{R}_{\text{boot}})^2} \tag{4.58}$$

is derived. From the available data points, B randomly assembled subsets are generated and for each subset the corresponding mean $\bar{R}_i$ is computed. Then, the mean of all these subset means $\bar{R}_{\text{boot}}$ can be determined. Equation (4.58) can be used to control the Monte Carlo method and serve as a stopping criteria, e.g. when the measurement uncertainty is undershot. Salomo et al. (2006) determined R with a precision of 2 nm. When applying bootstrapping Hesterberg (2015) pointed out to take certain precautions: $B > 15{,}000$ is required, narrowness bias needs to be identified and removed, and the distribution over $\bar{R}_i$ needs to be free of skewness.

To study the applicability of MCE_{boot} to the present method, the two distinctly different distributions depicted in Fig. 4.4 (a) and (d) are used. For both distributions, MCE_{mean} and MCE_{boot} are determined for an increasing number of data points. The results are shown in Fig. 4.5. The first and important result is that both MCE measures lead to similar curves confirming the exactness of MCE_{boot}. Furthermore, in additional studies an increase of B did not lead to higher deviation from MCE_{mean}. Therefore, in the following sections, the Monte Carlo method is considered to have converged when MCE_{boot} with $B = 15{,}000$ is less than some specified value, e.g. the experimental accuracy.

Secondly, the trivial observation can be made from Fig. 4.5 that a narrow distribution results in an overall lower error level. That means simulations of chains with boundary conditions and under external load converge faster than those of free chains.

[2] Figure reprinted with permission from Schulz et al. (2019). ©2019 by Elsevier

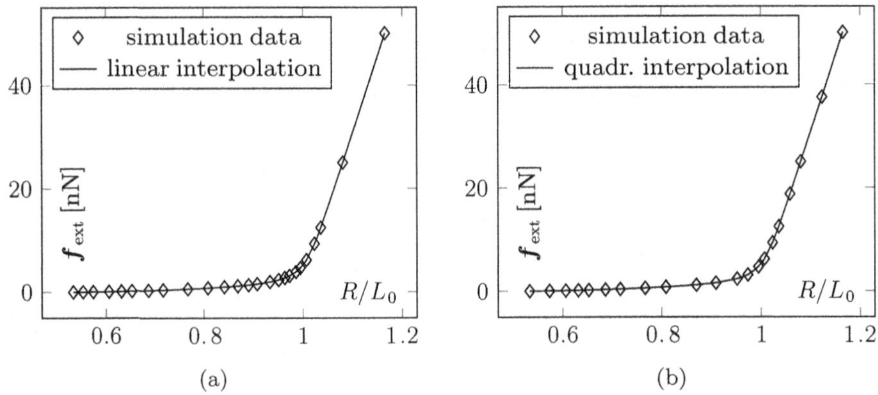

Fig. 4.6 Linear (**a**) and quadratic (**b**) interpolation between single data points from Monte Carlo simulations[3]

4.6.3 Obtaining Force-Extension Curves

To obtain a curve describing the force-extension relationship, a certain amount of data points are generated in a Monte Carlo manner and used as basis for interpolation. It is of course favorable to keep the basis as small as possible to reduce computation time. To this end, it is studied here how to intelligently select the position of the basis points.

The typical range of an experiment is from 0 to 50pN. At these points and in the middle at $f_{\text{ext}} = 25\text{pN}$ $\bar{R}$ is determined. At every further midpoint the value resulting from interpolation is compared to the Monte Carlo result and if the difference is less than the experimental accuracy no more data points are added in the interval. The results of this procedure are shown in Fig. 4.6 (a) and (b) with linear and quadratic interpolation, respectively. In these cases, 25 and 21 data points were required, respectively. Furthermore, it can be seen that smooth curves are generated conforming the correctness of MCE_{boot} as error measure.

4.6.4 Surrogate Modeling and Validation

In this section, the surrogate model based on a Gaussian process is set up and validated. First, the model parameters are chosen to be within a range such that 400 nm $\leq L_0 \leq$ 680 nm, 5 nm $\leq L_p \leq$ 200 nm, and $0.1\mu\text{N} \leq EA \leq 1\mu\text{N}$. These values were chosen based on the work of Salomo et al. (2006), who stretched a single DNA molecule and fitted the data using the worm-like chain model. The goal is that within this range the surrogate model produces the same values for the end-to-end distance

[3] Figure reprinted with permission from Schulz et al. (2019). ©2019 by Elsevier

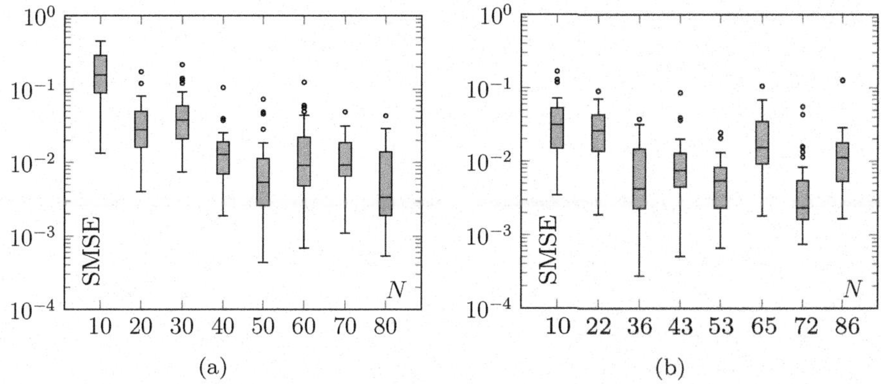

Fig. 4.7 The boxplots show the $SMSE$, see Eq. (4.59), for 30 validation points in dependence of the size N of the original (**a**) and reduced (**b**) training points. The *gray rectangle* contains half of all values between the *lower* and *upper quartile*. The *horizontal line* within that *rectangle* signifies the median. The *horizontal lines* outside the box symbolize the *lower* and *upper* whiskers stretching to 1.5 times the interquartile range in both directions. Outliers are depicted by *circles*[4]

as the original model using the procedure described in the last sections. For this end, two Latin hypercube samplings are performed filling this parameter space with n_{test} sample points having little correlation. The data values for $\boldsymbol{f}_{\text{ext}}$ in the training set are determined as described in the last section. In the validation set they are also generated according to the Latin hypercube method. One sampling set consisting of N points is used to train the Gaussian process and the other serves as validation set. The n_{test} sample points used for validation are denoted $\boldsymbol{x}_k$. Then the standardized mean squared error (Rasmussen & Williams, 2006)

$$SMSE = \frac{\sum_{k=1}^{n_{\text{test}}} ||\boldsymbol{g}(\boldsymbol{x}_k) - \boldsymbol{g}^*(\boldsymbol{x}_k)||}{STD} \tag{4.59}$$

measures the quality of the surrogate model, where $\boldsymbol{g}(\bullet)$ denotes the original Brownian dynamics model, $\boldsymbol{g}^*(\bullet)$ the Gaussian process, and STD is the standard deviation of all end-to-end lengths of a validation set. In Fig. 4.7 (a), $SMSE$ is plotted over the size N of the training data set. It can be seen that up to $N = 50$ $SMSE$ decreases and after that the median as well as the scattering fluctuate around 10^{-2}, which agrees well with the results published by Liu et al. (2018).

The more training data is used, the bigger the covariance matrix gets, which is detrimental to the computation speed since its inversion is required. To this end, it was tried to exclude sample points, which lead to simulation results not relevant for the regime of the experimental data. If a certain parameter combination results in an extension smaller than 90% or larger than 110% of the experimental extension,

[4] Figure reprinted with permission from Schulz et al. (2019). ©2019 by Elsevier

that combination was removed from the training data. Applying this procedure to the training set used to generate Fig. 4.7 (a), $SMSE$ starts to fluctuate around 10^{-2} above $N = 36$, see (b), and not just above $N = 50$. Based on these results, the parameter space is filled with 70 training points and based on the simulation results it is further reduced to at least 40 training points.

4.6.5 Fit to Experimental Data in a Bayesian Framework

With the hyperparameters of the surrogate model at hand, a fit to experimental data is performed in a Bayesian framework. This means, if $\boldsymbol{p}$ contains the material parameters, it is tried to solve Bayes' theorem reading

$$p(\boldsymbol{p}|\mathcal{D}) = \frac{p(\boldsymbol{p})p(\boldsymbol{y}|\boldsymbol{g}^*(\boldsymbol{X}), \boldsymbol{p})}{p(\boldsymbol{y}|\boldsymbol{g}^*(\boldsymbol{X}))}, \tag{4.60}$$

which is especially advantageous when limited experimental data is available. The prior $p(\boldsymbol{p})$ is chosen to be weakly informative, i.e. it is a normal distribution truncated according to the parameter interval, on which the surrogate model was trained. Additionally, the likelihood $p(\boldsymbol{y}|\boldsymbol{g}^*(\boldsymbol{X}), \boldsymbol{p})$ is a Gaussian, homoscedastic, and unknown noise term quantifying the disparity between the Gaussian process and the experimental results. Since the Gaussian process depends non-linearly on $\boldsymbol{p}$, numerical methods are required to find the posterior $p(\boldsymbol{p}|\mathcal{D})$. Here, the Stan-implementation (Stan Development Team, 2017) is used. It is a realization of the Hamiltonian Monte Carlo method and exhibits a fast convergence rate to the posterior (Carpenter et al., 2017). Only a brief description of the method is given. For more details the reader is referred to Gelman, Rubin, et al. (1992) and Stan Development Team (2017).

In the numerical scheme, four Markov chains are run. Then, the so-called potential scale reduction statistic value is computed. It is defined as the ratio of the variances of each chain to the combined variance of all four chains. If this value is less than 1.1 it indicates an overall convergence. Furthermore, to avoid autocorrelation among the generated quantities, the effective sample size is required to be larger than 1500.

Fit to experimental data from Salomo et al. (2006): The experimental data by Salomo et al. (2006) is fitted using the Stan-implementation. For details on the experimental setup see Sect. 4.6.1. A fit in the Bayesian framework results in a multivariate distribution over the parameters, which is presented and discussed in the following.

In Fig. 4.8 the one-dimensional distributions over the three single parameters are shown. It can be seen that L_0 shows a relatively narrow distribution. This means that based on the experimental data L_0 is less uncertain than the other parameters. In addition, the means and standard deviations of the parameter samplings are taken to plot the according Gaussian distributions in Fig. 4.8. All 3 sampling distributions appear to be non-symmetric about its mean and exhibit a slight tilt towards higher

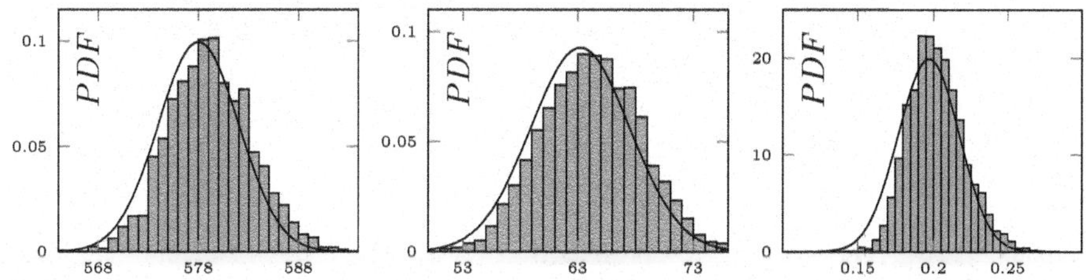

(a) L_0[nm] (b) L_p [nm] (c) EA [μN]

Fig. 4.8 The Bayesian fitting process to experimental data by Salomo et al. (2006) results in probability density functions (PDF) of the parameters (**a**) L_0, (**b**) L_p, and (**c**) EA, respectively. The mean and standard deviation of the parameter samplings are used to define and plot a Gaussian distribution[5]

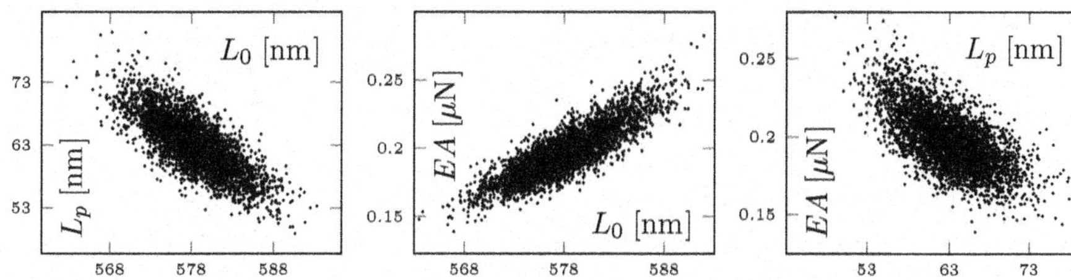

(a) (b) (c)

Fig. 4.9 The posterior over the parameters is a multivariate distribution whose marginalization over EA (**a**), L_p (**b**), and L_0 (**c**) are approximately depicted here using the results from the MCMC fitting procedure[6]

values. But a description of these distributions by mean and standard deviation seams accurate enough. The Bayesian inference does not result only in distributions over the single material parameters but also in a multi-variate (Gaussian) distribution over all material parameters. In the current application, a three-dimensional Gaussian is available. For better visualization, the three possible marginalizations resulting in two-dimensional distributions are depicted in Fig. 4.9. It can be seen that simple linear positive and negative correlations between the parameters are found. Figure 4.9a suggests that, based on the experimental data at hand, if a smaller value for L_p is chosen a smaller value for L_0 must be chosen as well to fit the data well.

The means and standard deviations of the parameter samplings are listed in Table 4.1 and compared to parameters obtained from other theoretical models. Marko & Siggia (1995) and Wang et al. (1997) derived statistical mechanics models for inextensible and extensible polymers, respectively. The model from Marko & Siggia

[5] Figure reprinted with permission from Schulz et al. (2019). ©2019 by Elsevier
[6] Figure reprinted with permission from Schulz et al. (2019). ©2019 by Elsevier

Table 4.1 The means and the standard deviations are listed for different models based on the same experimental data by Salomo et al. (2006)

Experiment	Theory	L_0[nm]	L_p[nm]	EA[pN]
Salomo et al. (2006)	present	578 ± 4	63.2 ± 4.3	197 ± 20
Salomo et al. (2006)	Marko & Siggia (1995)	652 ± 2	20.0 ± 5.0	–
Salomo et al. (2006)	Wang et al. (1997)	618 ± 1	34.0 ± 1.5	295 ± 10

(1995) is an interpolation formula of the exact solution based on a numeric solution scheme of entropic theory. It is applicable if the external force is less than $0.25(k_B T K_0^2/L_p)^{1/3}$. The parameters have similar physical meaning and can be carefully compared to the parameters of the present model. But it should be stressed that both models do not include any boundary conditions and no average end-to-end distance under absence of external forces is included. Moreover, Wang et al. (1997) simply added an enthalpic term to describe extensibility. Anyways, the model parameters resulting from the different strategies are roughly in the same range. It is found that L_0 is slightly smaller and L_p is distinctly larger compared to the theoretical models. The reasons for these differences most probably lie in the different physical assumptions and simplifications mentioned above.

Finally, in Fig. 4.10 the experimental data by Salomo et al. (2006) with error bars, the result of the Bayesian inference shown as gray band representing 90% of all curves, and the aforementioned theoretical models are depicted. In general, good agreement between the experiment and the new model is found. A big advantage of the current approach within the Bayesian framework is that the uncertainty of the target value is predicted by the model as well. The model for inextensible chains by Marko & Siggia (1995) fits only in the low-force regime. The Wang et al. (1997) model fits remarkably well despite the simplifications of the underlying model which gives rise to the question whether the quality of the fit is due to a good simplified physical model or due to the versatility of the equation used to fit the data.

Fit to experimental data from Wang et al. (1997): A second data set obtained by Wang et al. (1997) is fitted within the current Bayesian framework using the Stan-implementation (Stan Development Team, 2017). Wang et al. (1997) also performed experiments on double-stranded single molecules of DNA, the same molecule type that Salomo et al. (2006) used in their experiments.

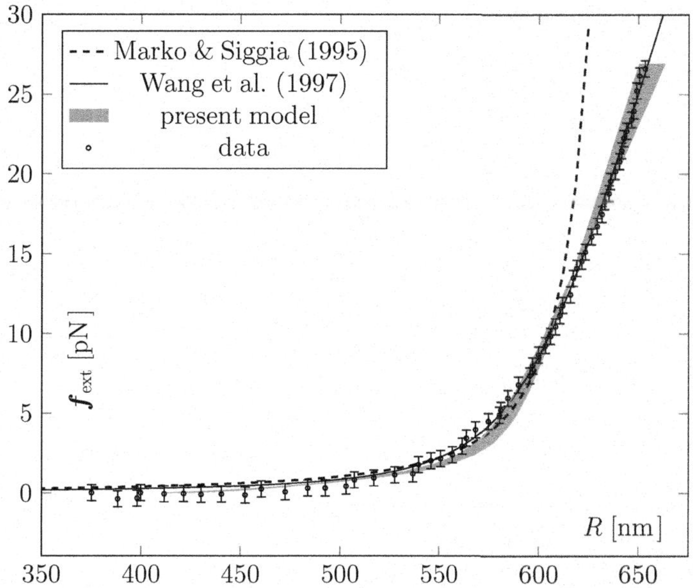

Fig. 4.10 Experimental data (Salomo et al., 2006) is shown with 95% uncertainty intervals. The result of the Bayesian inference is depicted as gray area representing 90% of all curves. Fits based on theoretical models by Marko & Siggia (1995) and Wang et al. (1997) are shown[7]

In the tests performed by Wang et al. (1997) one end of the DNA was chemically fixed to a cover glass and therefore modeled as fixed end. The other end was handled by an optical trap and was represented as hinged end in the simulation. The result of the Bayesian fitting procedure and the fit to the data with other models by Marko & Siggia (1995) and Wang et al. (1997) are shown in Fig. 4.11. The distributions over the material parameters resulting from the Bayesian fitting process are not illustrated for this data set since no new findings could be deduced. As found for the data by Salomo et al. (2006) the distributions over single material parameters can be adequately represented by Gaussian distributions. The corresponding means and standard deviations are listed in Table 4.2.

L_0 shows the narrowest distribution as it has the smallest variance in relation to its mean value. The same simple linear correlations between the three different material parameters were found as depicted in Fig. 4.9. Again, as was the case with the fit to the data by Salomo et al. (2006), using the new method, L_0 is smaller, L_p is larger, and EA is smaller compared to the other models. Furthermore, it is important to notice that even though in both experiments the same molecule type was used that

[7] Figure reprinted with permission from Schulz et al. (2019). ©2019 by Elsevier

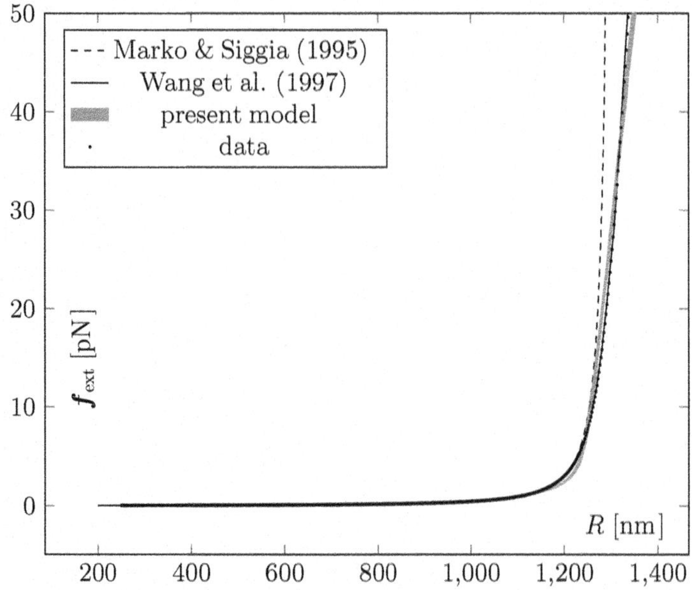

Fig. 4.11 The experimental data from Wang et al. (1997) is shown. The result of the Bayesian inference is depicted as *gray area* representing 90% of all *curves*. Furthermore, fits based on theoretical models by Marko & Siggia (1995) and Wang et al. (1997) are shown[8]

Table 4.2 The means and the standard deviations are listed for different models based on the same experimental data by Wang et al. (1997)

Experiment	Theory	L_0[nm]	L_p[nm]	EA[pN]
Wang et al. (1997)	present	1241 ± 2	87.5 ± 7.5	567 ± 23
Wang et al. (1997)	Marko & Siggia (1995)	1317 ± 2	43.5 ± 1.0	–
Wang et al. (1997)	Wang et al. (1997)	1314 ± 1	43.3 ± 0.5	1246 ± 10

all three compared models result in higher values for L_p and EA. The reasons for that remain unclear but possible causes might be the neglect of ionic repulsive forces and self-contact or underestimated inaccuracies of the experimental methods.

[8] Figures reprinted with permission from Schulz et al. (2019). ©2019 by Elsevier

Chapter 5
Conclusion

5.1 Novel Geometrically Exact Kirchhoff-Love Beams

A new Kirchhoff-Love beam theory was presented based on the strong enforcement of the Kirchhoff constraint by removing two translational degrees of freedom. Based on this theory two new geometrically exact Kirchhoff-Love beam elements, referred to as $\boldsymbol{\psi}^l$-element and $\boldsymbol{t}$-element, were devised, which differ in rotation parameterization and interpolation strategies. In general, the rotation tensor field and the axial strain field are discretized resulting in a weak form where the highest derivative is of first order and thus requiring C^0-continuous interpolation by Lagrangian polynomials of second order. 3-node elements with four nodal degrees of freedom are produced.

The $\boldsymbol{\psi}^l$-element employs the well-known local rotation vector (Jelenić & Crisfield, 1999) and treats the axial strain field as unknown resulting in a novel mixed Kirchhoff-Love finite-element formulation. This is in contrast to Zupan and Saje (2003, 2006), who interpolated all strain fields of a Simo-Reissner beam. The $\boldsymbol{\psi}^l$-element has the advantage of avoiding the computationally expensive determination of total rotations from the strains. Also, the present mixed formulation is not restricted to a straight undeformed configuration and was extended to a dynamic formulation, which would not be easily possible with the beam element presented by Ibrahimbegović et al. (1995), who also removed translational variables.

In the $\boldsymbol{t}$-element the tangent vector and a local twist parameter are the unknown quantities. For both elements, the objectivity, path-independence, and geometrical exactness were theoretically expected and numerically confirmed in appropriate examples. Arbitrary initial geometries and cross-sections can be realized. The $\boldsymbol{\psi}^l$-element does not exhibit numerical locking. In contrast, the $\boldsymbol{t}$-element shows locking effects for very slender beams, where the axial stiffness is higher than the bending rigidity. It would certainly be possible to remove this problem using any of the many techniques available in the literature, see e.g. Meier et al. (2015).

M. C. Schulz, *A New Kirchhoff-Love Beam Element and its Application to Polymer Mechanics*, Mechanics and Adaptronics,
https://doi.org/10.1007/978-3-031-06340-4_5

Further, the ψ^l-element was reformulated to obtain the ε-element, a reduced model, which strongly enforces the inextensibility of the center-line by simply removing the axial strain as unknown.

As could be expected, the ε-element showed the same results as the ψ^l-element and the element is not explicitly mentioned in the remainder of this section. In a numerical study, it was shown that the ψ^l-element is vastly superior compared to all other beam elements when only external moments act on the beam, which certainly has hardly any relevance in practice. When the beam is also loaded by external forces, a second and third order convergence rate were found for the tangent and displacement field, respectively, as theoretically expected. It is interesting to observe that on one hand, the integral to determine the center-line reduces the order of the interpolation, but on the other hand, the smoothing effect of the integral leads to an increase of the rate when the center-line is computed from the tangent. A study of the convergence order of the $\boldsymbol{t}$-element revealed the theoretically expected third and forth order reduction for the tangent and displacement field, respectively, independent of the nature of the load.

The discretization error of the ψ^l-element is comparable to that of other beam formulations by Jelenić and Crisfield (1999), Meier et al. (2014, 2015, 2017), and Simo (1985). The same holds for the $\boldsymbol{t}$-element with the additional condition that enough elements are used.

In a comparative study the required Newton iterations were documented and compared to the above mentioned beam formulations. The ψ^l-element uses the least iterations for the considered load cases whereas the $\boldsymbol{t}$-element shows poor performance for little element numbers. But when using more elements the performance of the elements by Meier et al. (2014, 2015) is reached.

A dynamic formulation of the ψ^l-element using the Generalized-α method for elements of $\mathbb{R}$ and $SO(3)$ was presented. An adequate conservation of energy was found in a numerical example. Nonetheless, the field of dynamic Kirchhoff-Love beam formulations appear to have room for improvement and constitute an interesting field of future research.

The novel elements are the first geometrically exact Kirchhoff-Love beams that can pass a patch test in an energetic and a geometrical sense.

The center-line of the new beam element is defined implicitly and needs to be computed via integration according to the Dirichlet boundary conditions. This leads to the drawback of the novel formulations that under certain conditions these boundary restrictions would require Lagrangian multipliers to be enforced. In addition, the implementation external loads is more involved, nested assembly steps are required, and the mass and damping matrices are not as sparse as in other formulations. Removing these shortcomings could be interesting topics for future research and might be realized by computing the center-line on element level, only. Another approach might be to assume constant axial strain in each element as done by Češarek et al. (2012). But this would limit the convergence order.

5.2 Finite-Element Brownian Dynamics Simulations for Bayesian Inference

The ψ^l-element, identified as advantageous in the first part of this thesis, is employed in the finite-element Brownian dynamics simulation established by Cyron and Wall (2009) where a polymer is modeled as a linearly elastic beam that is submerged in a Newtonian fluid and that experiences collisions with molecules of that fluid. To model these collisions a line load is applied, which contains a white noise approximation that is piece-wise constant in time and space and that is independent of the finite-element discretization. The emerging method was used to model whole force-extension experiments in a finite-element framework correctly incorporating the boundary conditions. In a Monte Carlo manner, mean values were obtained and adequately interpolated to obtain simulated force-extension curves with as little computational cost as possible. Further, it was demonstrated that the accuracy of the mean value determination process can be tracked using bootstrap statistics.

To further reduce computational cost, a surrogate model was set up, which directly gives the mean values of the Monte Carlo method based on the constitutive material parameters. To this end, a Gaussian process was trained on about 40 data sets. The surrogate model was used in a Bayesian framework to obtain the uncertainties and correlations of the material parameters for DNA based on experimental data by Salomo et al. (2006) and Wang et al. (1997). The found values were in the same order as those found in the literature. The chain length L_0 is slightly smaller and the persistence length L_p and the axial stiffness EA are larger. The reasons for that are hard to identify but probably have their basis in the different physical assumptions. The models used in the literature commonly disregard boundary conditions. Furthermore, unpublished simulations showed a dependence of the force-extension behavior on the friction coefficient. Thus, exploring new friction models or including the friction coefficient as unknown in the Bayesian parameter identification process is an interesting field for future research.

The presented numerical method constitutes a flexible framework for future enhancements to include different physical effects such as self-contact, electrostatic repulsion (Dobrynin, 2005), and entanglement with other polymer chains (Graham et al., 2003). Other thinkable extensions include more complex experiments, e.g. elongation in laminar flow or a (representative) network of interacting polymer chains (Kachan et al., 2016).

Bibliography

Abraham, R., Marsden, J. E., & Ratiu, T. (2012). *Manifolds, tensor analysis, and applications*. Springer.

Alder, B. J., & Wainwright, T. E. (1959). "Studies in molecular dynamics. I. General method". *The Journal of Chemical Physics, 31*(2), 459–466.

Alexander-Katz, A., Schneider, M. F., Schneider, S. W., Wixforth, A., & Netz, R. R. (2006). Shear-flow-induced unfolding of polymeric globules. *Physical review letters, 97*(13), 138101.

Allen, E. J., Novosel, S. J., & Zhang, Z. (1998). "Finite element and difference approximation of some linear stochastic partial differential equations". *Stochastics: An International Journal of Probability and Stochastic Processes, 64*(1–2), 117–142.

Antman, S. S. (1995). *Nonlinear problems of elasticity*. Springer.

Argyris, J. H., Balmer, H., Doltsinis, J. St., Dunne, P. C., Haase, M., Kleiber, M., Malejannakis, G. A., Mlejnek, H.-P., Müller, M., & Scharpf, D. W. (1979). Finite element method - The natural approach. *Computer Methods in Applied Mechanics and Engineering, 17*, 1–106.

Argyris, J. (1982). An excursion into large rotations. *Computer Methods in Applied Mechanics and Engineering, 32*(1–3), 85–155.

Armero, F., & Valverde, J. (2012). Invariant Hermitian finite elements for thin Kirchhoff rods. I: The linear plane case. *Computer Methods in Applied Mechanics and Engineering, 213–216*, 427–457.

Armero, F., & Valverde, J. (2012). Invariant Hermitian finite elements for thin Kirchhoff rods. II: The linear three-dimensional case. *Computer Methods in Applied Mechanics and Engineering, 213*, 458–485.

Arnold, M.,& Simeon, B. (2000). Pantograph and catenary dynamics: A benchmark problem and its numerical solution. *Applied Numerical Mathematics, 34*(4), 345–362.

Arnold, M., & Brüls, O. (2007). Convergence of the generalized-α scheme for constrained mechanical systems. *Multibody System Dynamics, 18*(2), 185–202.

Arnold, M., Cardona, A., & Brüls, O. (2016). "A Lie algebra approach to Lie Group Time Integration of Constrained Systems". *Structure-preserving Integrators in Nonlinear Structural Dynamics and Flexible Multibody Dynamics* (pp. 91–158). Springer International Publishing.

Arunakirinathar, K., & Reddy, B. D. (1993). Mixed finite element methods for elastic rods of arbitrary geometry. *Numerische Mathematik, 64*(1), 13–43.

Asher, M. J., Croke, B. F. W., Jakeman, A. J., & Peeters, L. J. M. (2015). A review of surrogate models and their application to groundwater modeling. *Water Resources Research, 51*(8), 5957–5973.

M. C. Schulz, *A New Kirchhoff-Love Beam Element and its Application to Polymer Mechanics*, Mechanics and Adaptronics,
https://doi.org/10.1007/978-3-031-06340-4

Aström, J. A., Kumar, P. S., Vattulainen, I., & Karttunen, M. (2008). Strain hardening, avalanches, and strain softening in dense cross-linked actin networks. *Physical Review E - Statistical, non-linear, and soft matter physics, 77*(5), 051913.
Aubram, D. (2009). *Differential geometry applied to continuum mechanics*. Shaker.
Baldo, G., Bonelli, A., Bursi, O., & Erlicher, S. (2006). The accuracy of the Generalized- α method in the time integration of non-linear Single-and Two-DOF forced systems. *Computational Mechanics, 38*(1), 15–31.
Barber, D. (2012). *Bayesian reasoning and machine learning*. Cambridge University Press.
Bathe, K.-J. (1982). *Finite element procedures in engineering analysis*. Prentice-Hall.
Bathe, K.-J. (2006). *Finite element procedures*. Klaus-Jürgen Bathe.
Bathe, K.-J., & Bolourchi, S. (1979). Large displacement analysis of three-dimensional beam structures. *International Journal for Numerical Methods in Engineering, 14*(7), 961–986.
Battini, J.-M., & Pacoste, C. (2002). Co-rotational beam elements with warping effects in instability problems. *Computer Methods in Applied Mechanics and Engineering, 191*(17–18), 1755–1789.
Bauchau, O. A., & Bottasso, C. L. (1999). On the design of energy preserving and decaying schemes for flexible, nonlinear multi-body systems. *Computer Methods in Applied Mechanics and Engineering, 169*(1), 61–80.
Bauer, A. M., Breitenberger, M., Philipp, B., Wüchner, R., & Bletzinger, K.-U. (2016). Nonlinear isogeometric spatial Bernoulli beam. *Computer Methods in Applied Mechanics and Engineering, 303*, 101–127.
Beatty, M. F. (1977). Vector analysis of finite rigid rotations. *Journal of Applied Mechanics, 44*(3), 501–502.
Belytschko, T., & Hsieh, B. J. (1973). Non-linear transient finite element analysis with convected co-ordinates. *International Journal for Numerical Methods in Engineering, 7*(3), 255–271.
Belytschko, T., & Glaum, L. W. (April 1979). Applications of higher order corotational stretch theories to nonlinear finite element analysis. *Computers and Structures, 10*(1–2), 175–182.
Belytschko, T., Liu, W. K., Moran, B., & Elkhodary, K. (2013). *Nonlinear finite elements for continua and structures*. John Wiley & Sons.
Bemis, J. E., Akhremitchev, B. B., & Walker, G. C. (1999). Single polymer chain elongation by atomic force microscopy. *Langmuir, 15*(8), 2799–2805.
Betsch, P., & Steinmann, P. (2002). Frame-indifferent beam finite elements based upon the geometrically exact beam theory. *International Journal for Numerical Methods in Engineering, 54*(12), 1775–1788.
Bishop, C. M. (2006). *Pattern recognition and machine learning*. springer.
Böl, M.,& Reese, S. (2005). Finite element modelling of rubber-like materials-a comparison between simulation and experiment. *Journal of materials science, 40*(22), 5933–5939.
Bonet A. J., Rubi, J. M., & Bedeaux, D. (1993). Dynamics of rodlike polymers in dilute solution. *Macromolecules, 26*(10), 2550–2561.
Borri, M., Ghiringhelli, G. L., & Merlini, T. (Jan. 1992). Linear analysis of naturally curved and twisted anisotropic beams. *Composites Engineering, 2*(5–7), 433–456.
de Borst, R., Crisfield, M. A., Remmers, J. J. C., & Verhoosel, C. V. (2012). *Nonlinear finite element analysis of solids and structures*. John Wiley & Sons.
Bottasso, C. L., & Borri, M. (1998). Integrating finite rotations. *Computer Methods in Applied Mechanics and Engineering, 164*(3–4), 307–331.
Bouchiat, C., Wang, M. D., Allemand, J. F., Strick, T., Block, S. M., & Croquette, V. (1999). Estimating the persistence length of a worm-like chain molecule from force-extension measurements. *Biophysical Journal, 76*(1), 409–413.
Boyer, F., & Primault, D. (2003). Finite element of slender beams in finite transformations: A geometrically exact approach. *International Journal for Numerical Methods in Engineering, 59*(5), 669–702.
Boyer, F., & Primault, D. (2004). Finite element of slender beams in finite transformations: A geometrically exact approach. *International Journal for Numerical Methods in Engineering, 59*(5), 669–702.

Brezzi, F., & Fortin, M. (1991). *Mixed and hybrid finite element methods, no. 15 in Springer Series in Computational Mathematics*. Springer.

Broersma, S. (1960). Rotational diffusion constant of a cylindrical particle. *The Journal of Chemical Physics, 32*(6), 1626–1631.

Brüls, O., & Cardona, A. (2010). On the use of Lie group time integrators in multibody dynamics. *Journal of Computational and Nonlinear Dynamics, 5*(3), 031002.

Brüls, O., Cardona, A.,& Arnold, M. (2012). Lie group generalized -α time integration of constrained flexible multibody systems. *Mechanism and Machine Theory, 48*, 121–137.

Buckdahn, R., & Pardoux, E. (1990). Monotonicity methods for white noise driven quasilinear SPDEs. *Diffusion processes and related problems in analysis, 1*, 219–233.

Cannarozzi, M., & Molari, L. (2008). A mixed stress model for linear elastodynamics of arbitrarily curved beams. *International Journal for Numerical Methods in Engineering, 74*(1), 116–137.

Cardona, A., & Geradin, M. (1988). A beam finite element non-linear theory with finite rotations. *International journal for numerical methods in engineering, 26*(11), 2403–2438.

Cardona, A., & Geradin, M. (1989). Time integration of the equations of motion in mechanism analysis. *Computers and Structures, 33*(3), 801–820.

Carpenter, B., Gelman, A., Hoffman, M. D., Lee, D., Goodrich, B., Betancourt, M., Brubaker, M. A., Guo, J., Li, P., & Riddell, A. (2017). Stan: A probabilistic programming language. *Grantee Submission, 76*(1), 1–32.

Češarek, P., Saje, M., & Zupan, D. (2012). Kinematically exact curved and twisted strain-based beam. *International Journal of Solids and Structures, 49*(13), 1802–1817.

Ceze, M., & Fidkowski, K. J. (2015). Constrained pseudo-transient continuation. *International Journal for Numerical Methods in Engineering, 102*(11), 1683–1703.

Chadwick, P. (2012). *Continuum mechanics: Concise theory and problems*. Courier Corporation.

Chandran, P. L., & Mofrad, M. R. K. (2010). Averaged implicit hydrodynamic model of semiflexible filaments. *Physical Review E, 81*(3), 031920.

Choi, J.-K., & Lim, J.-K. (1995). General curved beam elements based on the assumed strain fields. *Computers & structures, 55*(3), 379–386.

Chung, J., & Hulbert, G. M. (1993). A time integration algorithm for structural dynamics with improved numerical dissipation: The generalized -α method. *Journal of Applied Mechanics, 60*(2), 371–375.

Coffey, W. T., Kalmykov, Y. P., & Waldron, J. T. (2004). "Physics, Chemistry and Electrical Engineering".

Cook, R. D. (2007). *Concepts and applications of finite element analysis*. John Wiley & Sons.

Crisfield, M. A. (1983). An arc-length method including line searches and accelerations. *International Journal for Numerical Methods in Engineering, 19*(9), 1269–1289.

Crisfield, M. A. (1990). A consistent co-rotational formulation for non-linear, three-dimensional, beam-elements. *Computer Methods in Applied Mechanics and Engineering, 81*(2), 131–150.

Crisfield, M. A. (1997). *Non-linear finite element analysis of solids and structures* (Vol. 2). Wiley.

Crisfield, M. A., & Jelenić, G. (1999). Objectivity of strain measures in the geometrically exact three-dimensional beam theory and its finite-element implementation. *Proceedings of the Royal Society of London A: Mathematical, Physical and Engineering Sciences, 455*, 1125–1147.

Crisfield, M. A. (1981). A fast incremental/iterative solution procedure that handles 'snapthrough'. *Computers and Structures, 13*, 55–62.

Crivelli, L. A., & Felippa, C. A. (1993). A three-dimensional non-linear Timoshenko beam based on the core-congruential formulation. *International Journal for Numerical Methods in Engineering, 36*(21), 3647–3673.

Cyron, C. J., & Wall, W. A. (2012). Numerical method for the simulation of the Brownian dynamics of rod-like microstructures with three-dimensional nonlinear beam elements. *International Journal for Numerical Methods in Engineering, 90*(8), 955–987.

Cyron, C. J., Müller, K. W., Bausch, A. R., & Wall, W. A. (2013). Micromechanical simulations of biopolymer networks with finite elements. *Journal of Computational Physics, 244*, 236–251.

Cyron, C. J., & Wall, W. A. (2009). Finite-element approach to Brownian dynamics of polymers. *Physical Review E, 80*(6), 066704.

Cyron, C. J., & Wall, W. A. (2010). Consistent finite-element approach to Brownian polymer dynamics with anisotropic friction. *Physical Review E, 82*(6), 066705.

Demoures, F., Gay-Balmaz, F., Leyendecker, S., Ober-Blöbaum, S., Ratiu, T. S., & Weinand, Y. (2015). Discrete variational Lie group formulation of geometrically exact beam dynamics. *Numerische Mathematik, 130*(1), 73–123.

Deuflhard, P. (2005). *Newton methods for nonlinear problems: affine invariance and adaptive algorithms, 35*. Springer Science & Business Media

Dittmore, A., McIntosh, D. B., Halliday, S., & Saleh, O. A. (2011). Single-molecule elasticity measurements of the onset of excluded volume in poly (ethylene glycol). *Physical Review Letters, 107*(14), 148301.

Dobrynin, A. V. (2005). Electrostatic persistence length of semiflexible and flexible polyelectrolytes. *Macromolecules, 38*(22), 9304–9314.

Doi, & Edwards. (1988). *The theory of polymer dynamics* (Vol. 73). Oxford University Press.

Dokainish, M. A., & Subbaraj, K. (1989). "A survey of direct time-integration methods in computational structural dynamics-I. Explicit methods". *Computers and Structures, 32*(6), 1371–1386.

Drozdov, A. D., & Rabin, Y. (2000). "Elasticity of thin rods with spontaneous curvature and torsion-beyond geometrical lines". *arXiv preprint cond-mat/0002004*.

Durville, D. (2010). Simulation of the mechanical behaviour of woven fabrics at the scale of fibers. *International Journal of Material Forming, 3*(2), 1241–1251.

Duvenaud, D. (2014). "Automatic model construction with Gaussian processes". PhD thesis. University of Cambridge.

Eduardo, N. D., Eugenio, O., & Javier, O. (1988). On a non-linear formulation for curved Timoshenko beam elements considering large displacement/rotation increments. *International Journal for Numerical Methods in Engineering, 26*(7), 1597–1613.

Efron, B., & Tibshirani, R. J. (1994). *An introduction to the bootstrap*. CRC press.

Eldred, Giunta, & Collis. (2004). "Second-order corrections for surrogate-based optimization with model hierarchies". *10th AIAA/ISSMO multidisciplinary analysis and optimization conference*, 4457.

Elguedj, T., Gravouil, A., & Maigre, H. (2009). "An explicit dynamics extended finite element method. Part 1: Mass lumping for arbitrary enrichment functions". *Computer Methods in Applied Mechanics and Engineering, 198*(30–32), 2297–2317.

Erlicher, S., Bonaventura, L.,& Bursi, O. S. (2002). The analysis of the generalized- α method for non-linear dynamic problems. *Computational Mechanics, 28*(2), 83–104.

Ermak, D. L., & McCammon, J. A. (1978). Brownian dynamics with hydrodynamic interactions. *The Journal of chemical physics, 69*(4), 1352–1360.

Fixman, M., & Kovac, J. (1973). "Polymer conformational statistics. III. Modified Gaussian models of stiff chains". *The Journal of Chemical Physics, 58*(4), 1564–1568.

Flory, P. J. (1989). *Statistical mechanics of chain molecules*. Oxford University Press.

Freund, J., & Karakoç, A. (2016). Warping displacement of Timoshenko beam model. *International Journal of Solids and Structures, 92*, 9–16.

Fried, I., & Malkus, D. S. (1975). Finite element mass matrix lumping by numerical integration with no convergence rate loss. *International Journal of Solids and Structures, 11*(4), 461–466.

Frischkorn, J., & Reese, S. (2013). A solid-beam finite element and non-linear constitutive modelling. *Computer Methods in Applied Mechanics and Engineering, 265*, 195–212.

Gadot, B., Martinez, O. R., du Roscoat, S. R., Bouvard, D., Rodney, D., & Orgéas, L. (2015). Entangled single-wire NiTi material: A porous metal with tunable superelastic and shape memory properties. *Acta Materialia, 96*, 311–323.

Gee, Michael W., Kelley, C. T., & Lehoucq, R. B. (2009). Pseudo-transient continuation for nonlinear transient elasticity. *International Journal for Numerical Methods in Engineering, 78*(10), 1209–1219.

Gelman, A., Rubin, D. B., & et al. (1992). Inference from iterative simulation using multiple sequences. *Statistical science, 7*(4), 457–472.

Ghosh, S., & Roy, D. (2008). Consistent quaternion interpolation for objective finite element approximation of geometrically exact beam. *Computer Methods in Applied Mechanics and Engineering, 198*(3–4), 555–571.

Ghosh, S., & Roy, D. (2009). A frame-invariant scheme for the geometrically exact beam using rotation vector parametrization. *Computational Mechanics, 44*(1), 103.

Giavotto, V., Borri, M., Mantegazza, P., Ghiringhelli, G., Carmaschi, V., Maffioli, G. C., & Mussi, F. (1983). Anisotropic beam theory and applications. *Computers and Structures, 16*(1–4), 403–413.

Gill, P. E., & Murray, W. (1974). Safeguarded steplength algorithms for optimization using descent methods. *National Physical Laboratory, Division of Numerical Analysis and Computing, 37*, 1–37.

Gonzalez, O. (1996). Time integration and discrete Hamiltonian systems. *Journal of Nonlinear Science, 6*(5), 449.

Gooneie, A., Schuschnigg, S., & Holzer, C. (2017). A review of multiscale computational methods in polymeric materials. *Polymers, 9*(1), 16.

Graham, R. S., Likhtman, A. E., McLeish, T. C. B., & Milner, S. T. (2003). Microscopic theory of linear, entangled polymer chains under rapid deformation including chain stretch and convective constraint release. *Journal of Rheology, 47*(5), 1171–1200.

Greco, L., & Cuomo, M. (2016). An isogeometric implicit G1 mixed finite element for Kirchhoff space rods. *Computer Methods in Applied Mechanics and Engineering, 298*, 325–349.

Greco, L., & Cuomo, M. (2013). B-Spline interpolation of Kirchhoff-Love space rods. *Computer Methods in Applied Mechanics and Engineering, 256*, 251–269.

Gurtin, M. E. (1982). *An introduction to continuum mechanics* (Vol. 158). Academic press.

Hallatschek, O., Frey, E.,& Kroy, K. (2007). "Tension dynamics in semiflexible polymers. I. Coarsegrained equations of motion". *Physical Review E, 75*(3), 031905.

Han, S., & Bauchau, O. A. (November 2016). Nonlinear, three-dimensional beam theory for dynamic analysis. *Multibody System Dynamics, 41*(2), 173–200.

Hegemier, G. A., & Nair, S. (1977). A nonlinear dynamical theory for heterogeneous, anisotropic, elasticrods. *AIAA Journal, 15*(1), 8–15.

Hellinger, E. (1907). *Die Allgemeinen Ansaetze der Mechanik der Kontinua* (pp. 601-694). Vieweg+Teubner Verlag.

Hestenes, M. R., & Stiefel, E. (1952). *Methods of conjugate gradients for solving linear systems* (Vol. 49). NBS.

Hesterberg, T. C. (2015). What teachers should know about the bootstrap: Resampling in the undergraduate statistics curriculum. *The American Statistician, 69*(4), 371–386.

Heussinger, C., & Frey, E. (2007). Force distributions and force chains in random stiff fiber networks. *The European Physical Journal E - Soft Matter, 24*(1), 47–53.

Hilber, H. M., Hughes, J. R. T., & Taylor, R. L. (1977). Improved numerical dissipation for time integration algorithms in structural dynamics. *Earthquake Engineering and Structural Dynamics, 5*(3), 283–292.

Hinton, E., Rock, T., & Zienkiewicz, O. C. (1976). A note on mass lumping and related processes in the finite element method. *Earthquake Engineering and Structural Dynamics, 4*(3), 245–249.

Hjelmstad, K. D. (2007). *Fundamentals of structural mechanics*. Springer Science & Business Media.

Hodges, D. H. (1990). A mixed variational formulation based on exact intrinsic equations for dynamics of moving beams. *International journal of solids and structures, 26*(11), 1253–1273.

Hodges, D. H. (2003). Geometrically exact, intrinsic theory for dynamics of curved and twisted anisotropic beams. *AIAA Journal, 41*(6), 1131–1137.

Holzapfel, G. A., & Ogden, R. W. (2011). On the bending and stretching elasticity of biopolymer filaments. *Journal of Elasticity, 104*(1–2), 319–342.

Hu, H. C. (1955). On some variational methods on the theory of elasticity and the theory of plasticity. *Scientia Sinica, 4*(1), 33–54.

Hugel, T., Grosholz, M., Clausen-Schaumann, H., Pfau, A., Gaub, H., & Seitz, M. (2001). Elasticity of single polyelectrolyte chains and their desorption from solid supports studied by AFM based single molecule force spectroscopy. *Macromolecules, 34*(4), 1039–1047.

Hughes, J. R. T. (2012). *The finite element method: Linear static and dynamic finite element analysis.* Courier Corporation.

Hulbert, G. M. (2017). "Computational structural dynamics". *Encyclopedia of Computational Mechanics Second Edition,* 1-25.

Hulbert, G. M., & Chung, J. (1996). Explicit time integration algorithms for structural dynamics with optimal numerical dissipation. *Computer Methods in Applied Mechanics and Engineering, 137*(2), 175–188.

Ibrahimbegovic, A. (1997). On the choice of finite rotation parameters. *Computer Methods in Applied Mechanics and Engineering, 149*(1–4), 49–71.

Ibrahimbegovic, A. M., & Mamouri, S. (2002). Energy conserving/decaying implicit time-stepping scheme for nonlinear dynamics of three-dimensional beams undergoing finite rotations. *Computer Methods in Applied Mechanics and Engineering, 191*(37–38), 4241–4258.

Ibrahimbegović, A. (1995). On finite element implementation of geometrically nonlinear Reissner's beam theory: Three-dimensional curved beam elements. *Computer Methods in Applied Mechanics and Engineering, 122*(1–2), 11–26.

Ibrahimbegović, A., Frey, F., & Kožar, I. (1995). Computational aspects of vector-like parametrization of three-dimensional finite rotations. *International Journal for Numerical Methods in Engineering, 38*(21), 3653–3673.

Iura, M., & Atluri, S. N. (1988). Dynamic analysis of finitely stretched and rotated three-dimensional space-curved beams. *Computers and structures, 29*(5), 875–889.

Iura, M., & Atluri, S. N. (1988). On a consistent theory, and variational formulation of finitely stretched and rotated 3-D space-curved beams. *Computational Mechanics, 4*(2), 73–88.

Jafari, M., & Mahjoob, M. J. (August 2010). An exact three-dimensional beam element with nonuniform cross section. *Journal of Applied Mechanics, 77,* 6.

Jelenić, G., & Crisfield, M. A. (1998). Interpolation of rotational variables in nonlinear dynamics of 3D beams. *International Journal for Numerical Methods in Engineering, 43*(7), 1193–1222.

Jelenić, G., & Crisfield, M. A. (1999). Geometrically exact 3D beam theory: Implementation of a strain-invariant finite element for statics and dynamics. *Computer Methods in Applied Mechanics and Engineering, 171*(1–2), 141–171.

Jelenić, G., & Saje, M. (1995). A kinematically exact space finite strain beam model - finite element formulation by generalized virtual work principle. *Computer Methods in Applied Mechanics and Engineering, 120*(1–2), 131–161.

Ji, X. Y., & Feng, X. Q. (2011). Coarse-grained mechanochemical model for simulating the dynamic behavior of microtubules. *Physical Review E - Statistical, nonlinear, and soft matter physics, 84*(3), 031933.

Kachan, D., Müller, K. W., Wall, W. A., & Levine, A. J. (2016). Discontinuous bundling transition in semiflexible polymer networks induced by Casimir interactions. *Physical Review E, 94*(3), 032505.

Kane, C., Marsden, J. E., Ortiz, M., & West, M. (2000). Variational integrators and the Newmark algorithm for conservative and dissipative mechanical systems. *International Journal for Numerical Methods in Engineering, 49*(10), 1295–1325.

Kapania, R. K., & Li, J. (2003). A formulation and implementation of geometrically exact curved beam elements incorporating finite strains and finite rotations. *Computational Mechanics, 30*(5–6), 444–459.

Karamanlidis, D., & Jasti, R. (1987). Curved mixed beam elements for the analysis of thinwalled free-form arches. *Ingenieur-Archiv, 57*(5), 361–367.

Karatrantos, A., Clarke, N., & Kröger, M. (2016). Modeling of polymer structure and conformations in polymer nanocomposites from atomistic to mesoscale: A Review. *Polymer Reviews, 56*(3), 385–428.

Kasas, S., Cibert, C., Kis, A., De Los Rios, P., Riederer, B. M., Forro, L., Dietler, G., & Catsicas, S. (2004). Oscillation modes of microtubules. *Biology of the Cell, 96*(9), 697–700.

Kasas, S., Kis, A., Riederer, B. M., Forró, L., Dietler, G., & Catsicas, S. (2004). Mechanical properties of microtubules explored using the finite elements method. *ChemPhysChem, 5*(2), 252–257.

Kelley, C. T. (2003). *Solving nonlinear equations with Newton's method* (Vol. 1). SIAM.

Kelley, C. T., & Keyes, D. E. (1998). Convergence analysis of pseudo-transient continuation. *SIAM Journal on Numerical Analysis, 35*(2), 508–523.

Kirchhoff, G. (1859). Ueber das Gleichgewicht und die Bewegung eines unendlich dünnen elastischen Stabes. *Journal für die reine und angewandte Mathematik, 56*, 258–313.

Klumpp, A. R. (1976). Singularity-free extraction of a quaternion from a direction-cosine matrix. *Journal of Spacecraft and Rockets, 13*(12), 754–755.

Koehler, E., and Brown, E., & Haneuse, S. J.-P. A. (2009). On the assessment of Monte Carlo error in simulationbased statistical analyses. *The American Statistician, 63*(2), 155–162.

Kondoh, K., Tanaka, K., & Atluri, S. N. (1986). An explicit expression for the tangent-stiffness of a finitely deformed 3-D beam and its use in the analysis of space frames. *Computers & structures, 24*(2), 253–271.

Kulachenko, A., Denoyelle, T., Galland, S., & Lindström, S. B. (March 2012). Elastic properties of cellulose nanopaper. *Cellulose, 19*(3), 793–807.

Le Goff, L., Hallatschek, O., Frey, E., & Amblard, F. (2002). Tracer studies on f-actin fluctuations. *Physical Review Letters, 89*(25), 258101.

Lee, P.-G., & Sin, H.-C. (1993). Locking-free straight beam element based on curvature. *Communications in numerical methods in engineering, 9*(12), 1005–1011.

Lee, S. H. (1989). Rudimentary considerations for effective line search method in nonlinear finite element analysis. *Computers and Structures, 32*(6), 1287–1301.

Lens, E. V., & Cardona, A. (2008). A nonlinear beam element formulation in the framework of an energy preserving time integration scheme for constrained multibody systems dynamics. *Computers and Structures, 86*(1–2), 47–63.

Lieleg, O., Schmoller, K. M., Cyron, C. J., Luan, Y., Wall, W. A., and Bausch, A. R. (2009). Structural polymorphism in heterogeneous cytoskeletal networks. *Soft Matter, 5*(9), 1796–1803.

Lin, Y., Wei, X., Qian, J., Sze, K. Y., & Shenoy, V. B. (2014). A combined finite element-Langevin dynamics (FEM-LD) approach for analyzing the mechanical response of bio-polymer networks. *Journal of the Mechanics and Physics of Solids, 62*, 2–18.

Linke, W. A., Kulke, M., Li, H., Fujita-Becker, S., Neagoe, C., Manstein, D. J., Gautel, M., & Fernandez, J. M. (2002). PEVK domain of titin: An entropic spring with actin-binding properties. *Journal of Structural Biology, 137*(1), 194–205.

Liu, H., Cai, J., & Ong, Y.-S. (2018). Remarks on multi-output Gaussian process regression. *Knowledge-Based Systems, 144*, 102–121.

Liu, X., & Pollack, G. H. (2002). Mechanics of F-actin characterized with microfabricated cantilevers. *Biophysical Journal, 83*(5), 2705–2715.

Liu, X., Zhu, Q., & Lu, H. (2014). Modeling multiresponse surfaces for airfoil design with multipleoutput- Gaussian-process regression. *Journal of Aircraft, 51*(3), 740–747.

Love, A. E. H. (1944). *A treatise on the mathematical theory of elasticity*. Courier Corporation.

Love, A. E. H. (2013). *A treatise on the mathematical theory of elasticity*. Cambridge university press.

Lyly, M., Stenberg, R., & Vihinen, T. (1993). A stable bilinear element for the Reissner-Mindlin plate model. *Computer Methods in Applied Mechanics and Engineering, 110*(3–4), 343–357.

MacKintosh, F. C. (2006). Polymer-based models of cytoskeletal networks. *Cytoskeletal Mechanics: Models and Measurements in Cell Mechanics, 9*(3), 152–169.

Mai, D. J., Saadat, A., Khomami, B., & Schroeder, C. M. (2018). Stretching dynamics of single comb polymers in extensional flow. *Macromolecules, 51*(4), 1507–1517.

Mäkinen, J. (2007). Total Lagrangian Reissner's geometrically exact beam element without singularities. *International Journal for Numerical Methods in Engineering, 70*(9), 1009–1048.

Manca, F., Giordano, S., Palla, P. L., Zucca, R., Cleri, F., & Colombo, L. (2012). Elasticity of flexible and semiflexible polymers with extensible bonds in the Gibbs and Helmholtz ensembles. *The Journal of Chemical Physics, 136*(15), 154906.

Mark, J. E. (2007). *Physical properties of polymers handbook.* Springer.

Marko, J. F., & Siggia, E. D. (1995). Stretching dna. *Macromolecules, 28*(26), 8759–8770.

Mata, P., Oller, S., & Barbat, A. H. (2007). Static analysis of beam structures under nonlinear geometric and constitutive behavior. *Computer methods in applied mechanics and engineering, 196*(45–48), 4458–4478.

Meier, C., Popp, A., & Wall, W. A. (2014). An objective 3D large deformation finite element formulation for geometrically exact curved Kirchhoff rods. *Computer Methods in Applied Mechanics and Engineering, 278*, 445–478.

Meier, C., Popp, A, & Wall, W. A. (2015). A locking-free finite element formulation and reduced models for geometrically exact Kirchhoff rods. *Computer Methods in Applied Mechanics and Engineering, 290*, 314–341.

Meier, C. Popp, A., & Wall, W. A. (2017). Geometrically Exact Finite Element Formulations for Slender Beams: Kirchhoff-Love Theory Versus Simo-Reissner Theory. *Archives of Computational Methods in Engineering, 26*(1), 163–243.

Montesi, A., Morse, D. C., & Pasquali, M. (2005). Brownian dynamics algorithm for bead-rod semiflexible chain with anisotropic friction. *The Journal of chemical physics, 122*(8), 084903.

Müller, K, W., Meier, C., & Wall, W. A. (2015). Resolution of sub-element length scales in Brownian dynamics simulations of biopolymer networks with geometrically exact beam finite elements. *Journal of Computational Physics, 303*, 185–202.

Munson, K. M., Mulugeta, P. G., & Donhauser, Z. J. (2007). Enhanced mechanical stability of microtubules polymerized with a slowly hydrolyzable nucleotide analogue. *The Journal of Physical Chemistry B, 111*(19), 5053–5057.

Netz, R. R. (2001). Strongly stretched semiflexible extensible polyelectrolytes and DNA. *Macromolecules, 34*(21), 7522–7529.

Newmark, N. M. (1959). A method of computation for structural dynamics. *Journal of the Engineering Mechanics Division, 85*(3), 67–94.

Nocedal, J., & Wright, S. J. (1999). *Numerical optimization, 35*(67–68), 7.

Noll, W. (2011). *Non-linear Continuum Theories* (pp. 159–200). Springer.

Noor, A. K., & Peters, J. M. (1981). Mixed models and reduced/selective integration displacement models for nonlinear analysis of curved beams. *International Journal for Numerical Methods in Engineering, 17*(4), 615–631.

Noor, A. K., & Peters, J. M. (1981). Mixed models and reduced/selective integration displacement models for nonlinear analysis of curved beams. *International Journal for Numerical Methods in Engineering, 17*(4), 615–631.

Nukala, P. K. V. V., & White, D. W. (2004). A mixed finite element for three-dimensional nonlinear analysis of steel frames. *Computer Methods in Applied Mechanics and Engineering, 193*(23–26), 2507–2545.

Ortega, A., & Garcıa de la Torre, J. (2003). Hydrodynamic properties of rodlike and disklike particles in dilute solution. *The Journal of chemical physics, 119*(18), 9914–9919.

Ortega, J M., & Rheinboldt, W. C. (1970). *Iterative solution of nonlinear equations in several variables* (Vol. 30). SIAM.

Park, J. D., Myung, J. S., & Ahn, K. H. (2016). A review on particle dynamics simulation techniques for colloidal dispersions: Methods and applications. *Korean Journal of Chemical Engineering, 33*(11), 3069–3078.

Park, M. S., & Lee, B. C. (1996). Geometrically non-linear and elastoplastic three-dimensional shear flexible beam element of von-Mises-type hardening material. *International Journal for Numerical Methods in Engineering, 39*(3), 383–408.

Petrov, E., & Géradin, M. (1998). Finite element theory for curved and twisted beams based on exact solutions for three-dimensional solids Part 1: Beam concept and geometrically exact nonlinear formulation. *Computer Methods in Applied Mechanics and Engineering, 165*(1–4), 43–92.

Pimenta, P. M., & Yojo, T. (1993). Geometrically exact analysis of spatial frames. *Applied Mechanics Reviews, 46*(11S), 118–128.

Pittman, C. U., & Culbertson, B. M. (2012). *New monomers and polymers*. Springer Science & Business Media.

Poelert, S. L., Weinans, H. H., & Zadpoor, A. A. (2011). "Thermal Fluctuations of Anisotropic Semiflexible Polymers". arXiv preprint arXiv:1110.5613.

Prathap, G. (1985). The curved beam/deep arch/finite ring element revisited. *International Journal for Numerical Methods in Engineering, 21*(3), 389–407.

Prathap, G., & Naganarayana, B. P. (1990). Analysis of locking and stress oscillations in a general curved beam element. *International Journal for Numerical Methods in Engineering, 30*(1), 177–200.

Purohit, P. K., Arsenault, M. E., Goldman, Y., & Bau, H. H. (2008). The mechanics of short rod-like molecules in tension. *International Journal of Non-Linear Mechanics, 43*(10), 1056–1063.

Arnold, M., & Simeon, B. (2010). *Numerical mathematics* (Vol. 37). Springer Science & Business Media.

Rahman, A. (1964). Correlations in the motion of atoms in liquid argon. *Physical Review, 136*(2A), A405.

Rasmussen, C. E., & Williams, C. K. I. (2006). *Gaussian processes for machine learning*. MIT Press.

Reddy, B. D. (2013). *Introductory functional analysis: With applications to boundary value problems and finite elements* (Vol. 27). Springer Science & Business Media.

Reddy, J. N. (2014). *An introduction to nonlinear finite element analysis: With applications to heat transfer, fluid mechanics, and solid mechanics*. OUP Oxford.

Reissner, E. (1981). On finite deformations of space-curved beams. *Zeitschrift für angewandte Mathematik und Physik ZAMP, 32*(6), 734–744.

Reissner, E. (1950). On a variational theorem in elasticity. *Journal of Mathematics and Physics, 29*(1–4), 90–95.

Reissner, E. (1972). On one-dimensional finite-strain beam theory: the plane problem. *Zeitschrift für angewandte Mathematik und Physik ZAMP, 23*(5), 795–804.

Ritto-Corrêa, M, & Camotim, D. (2003). Work-conjugacy between rotation-dependent moments and finite rotations. *International journal of solids and structures, 40*(11), 2851–2873.

Ritto-Corrêa, M., & Camotim, D. (2008). On the arc-length and other quadratic control methods: Established, less known and new implementation procedures. *Computers and Structures, 86*(11–12), 1353–1368.

Robert, C., & Casella, J. (2013). *Monte Carlo statistical methods*. Springer Science & Business Media.

Robinson, T. D., Eldred, M. S., Willcox, K. E., & Haimes, R. (2008). Surrogate-based optimization using multifidelity models with variable parameterization and corrected space mapping. *AIAA Journal, 46*(11), 2814–2822.

Romero, I., & Armero, F. (2002). An objective finite element approximation of the kinematics of geometrically exact rods and its use in the formulation of an energy-momentum conserving scheme in dynamics. *International Journal for Numerical Methods in Engineering, 54*(12), 1683–1716.

Romero, I., Urrecha, M., & Cyron, C. J. (2014). A torsion-free non-linear beam model. *International Journal of Non-Linear Mechanics, 58*, 1–10.

Romero, I. (2004). The interpolation of rotations and its application to finite element models of geometrically exact rods. *Computational Mechanics, 34*(2), 121–133.

Romero, I. (2008). A comparison of finite elements for nonlinear beams: The absolute nodal coordinate and geometrically exact formulations. *Multibody System Dynamics, 20*(1), 51–68.

Romero, I. (2008). Formulation and performance of variational integrators for rotating bodies. *Computational Mechanics, 42*(6), 825–836.

Rossmann, W. (2002). *Lie groups: An introduction through linear groups* (Vol. 5). Oxford University Press on Demand.

Rotne, J., & Prager, S. (1969). Variational treatment of hydrodynamic interaction in polymers. *The Journal of Chemical Physics, 50*(11), 4831–4837.

Rubinstein, M., & Colby, R. H. (2003). *Polymer Physics*. Oxford University Press.

Saad, Y. (2003). *Iterative methods for sparse linear systems* (Vol. 82). SIAM.

Saitô, N., Takahashi, K., & Yunoki, Y. (1967). The statistical mechanical theory of stiff chains. *Journal of the Physical Society of Japan, 22*(1), 219–226.

Salomo, M., Kegler, K., Gutsche, C., Struhalla, M., Reinmuth, J., Skokow, W., Hahn, U., & Kremer, F. (2006). The elastic properties of single double-stranded DNA chains of different lengths as measured with optical tweezers. *Colloid and Polymer Science, 284*(11), 1325–1331.

Sander, O. (2010). Geodesic finite elements for Cosserat rods. *International Journal for Numerical Methods in Engineering, 82*(13), 1645–1670.

Sansour, C., Sansour, J., & Wriggers, P. (1996). A finite element approach to the chaotic motion of geometrically exact rods undergoing in-plane deformations. *Nonlinear Dynamics, 11*(2), 189–212.

Sansour, C., & Wagner, W. (2003). Multiplicative updating of the rotation tensor in the finite element analysis of rods and shells-a path independent approach. *Computational Mechanics, 31*(1–2), 153–162.

Santos, H. A. F. A., Pimenta, P. M., & De Almeida, J. P. M. (2010). Hybrid and multi-field variational principles for geometrically exact three-dimensional beams. *International Journal of Non-Linear Mechanics, 45*(8), 809–820.

Schulz, M., & Böl, M. (2019). A finite element formulation for a geometrically exact Kirchhoff-Love beam based on constrained translation. *Computational Mechanics, 64*(4), 1155–1175.

Schulz, M., Dittmann, J., & Böl, M. (2019). Modeling the mechanical behavior of semi- flexible polymer chains using a surrogate model based on a finite-element approach to Brownian polymer dynamics. *Journal of the Mechanics and Physics of Solids, 130*, 101–117.

Shi, Y., He, S., & Hearst, J. E. (1996). Statistical mechanics of the extensible and shearable elastic rod and of DNA. *The Journal of Chemical Physics, 105*(2), 714–731.

Shi, Y., Borovik, A. E., & Hearst, J. E. (1995). Elastic rod model incorporating shear and extension, generalized nonlinear Schrödinger equations, and novel closed-form solutions for supercoiled DNA. *The Journal of Chemical Physics, 103*(8), 3166–3183.

Simo, J. C. (1985). "A finite strain beam formulation. The three-dimensional dynamic problem. Part I". *Computer Methods in Applied Mechanics and Engineering, 49*(1), 55–70.

Simo, J. C., & Fox, D. D. (1989). "On a stress resultant geometrically exact shell model. Part I: Formulation and optimal parametrization". *Computer Methods in Applied Mechanics and Engineering, 72*(3), 267–304.

Simo, J. C., Fox, D. D., & Rifai, M. S. (1990). "On a stress resultant geometrically exact shell model. Part III: Computational aspects of the nonlinear theory". *Computer Methods in Applied Mechanics and Engineering, 79*(1), 21–70.

Simo, J. C., and Vu-Quoc, L. (1986). "A three-dimensional finite-strain rod model. Part II: Computational aspects". *Computer Methods in Applied Mechanics and Engineering, 58*(1), 79–116.

Simo, J. C., and Vu-Quoc, L. (1988). On the dynamics in space of rods undergoing large motions-a geometrically exact approach. *Computer Methods in Applied Mechanics and Engineering, 66*(2), 125–161.

Simo, J. C., & Wong, K. K. (1991). Unconditionally stable algorithms for rigid body dynamics that exactly preserve energy and momentum. *International Journal for Numerical Methods in Engineering, 31*(1), 19–52.

Smith, S. B., Finzi, L., & Bustamante, C. (1992). Direct mechanical measurements of the elasticity of single DNA molecules by using magnetic beads. *Science, 258*(5085), 1122–1126.

Sonneville, V., Cardona, A., & Brüls, O. (2014). Geometrically exact beam finite element formulated on the special Euclidean group SE (3). *Computer Methods in Applied Mechanics and Engineering, 268*, 451–474.

Sonneville, V., Cardona, A., & Brüls, O. (2014). Geometrically exact beam finite element formulated on the special Euclidean group SE (3). *Computer Methods in Applied Mechanics and Engineering, 268*, 451–474.

Sotoudeh, Z., & Hodges, D. H. (2011). Modeling beams with various boundary conditions using fully intrinsic equations. *Journal of Applied Mechanics, 78*(3), 031010.

Spurrier, R. A. (1978). Comment on singularity-free extraction of a quaternion from a direction-cosine matrix. *Journal of Spacecraft and Rockets, 15*(4), 255.

Stan Development Team. (2017). Stan Modeling Language Users Guide and Reference Manual, Version 2.17. https://mc-stan.org

Stolarski, H., & Belytschko, T. (1982). "Membrane locking and reduced integration for curved elements".

Strang, G., and Fix, G. J. (2008). *An analysis of the finite element method* (Vol. 212). Wellesley-Cambridge Press.

Strick, T. R., Dessinges, M. N., Charvin, G., Dekker, N. H., Allemand, J. F., Bensimon, D., & Croquette, V. (2002). Stretching of macromolecules and proteins. *Reports on Progress in Physics, 66*(1), 1.

Stuelpnagel, J. (1964). On the parametrization of the three-dimensional rotation group. *SIAM Review, 6*(4), 422–430.

Subbaraj, K., & Dokainish, M. A. (1989). "A survey of direct time-integration methods in computational structural dynamics-II. Implicit methods". *Computers and Structures, 32*(6), 1387–1401.

un, Y.-L., Luo, Z.-P., Fertala, A., & An, K.-N. (2002). Direct quantification of the flexibility of type I collagen monomer. *Biochemical and Biophysical Research Communications, 295*(2), 382–386.

Teh, L. H., & Clarke, M. J. (1997). New definition of conservative internal moments in space frames. *Journal of Engineering Mechanics, 123*(2), 97–106.

Tessler, A., & Spiridigliozzi, L. (1986). Curved beam elements with penalty relaxation. *International Journal for Numerical Methods in Engineering, 23*(12), 2245–2262.

Tian, T., & Burrage, K. (2001). Implicit Taylor methods for stiff stochastic differential equations. *Applied Numerical Mathematics, 38*(1–2), 167–185.

Tserpes, K. I., & Papanikos, P. (2005). Finite element modeling of single-walled carbon nanotubes. *Composites Part B: Engineering, 36*(5), 468–477.

van Dillen, T., Onck, P. R., & Van der Giessen, E. (2006). "Stiffening of semiflexible biopolymers and cross-linked networks". *ArXiv Physics e-prints.*

Wackerfuß, J., & Gruttmann, F. (2009). A mixed hybrid finite beam element with an interface to arbitrary three-dimensional material models. *Computer Methods in Applied Mechanics and Engineering, 198*(27–29), 2053–2066.

Wang, M. D., Yin, H., Landick, R., Gelles, J., & Block, S. M. (1997). Stretching DNA with optical tweezers. *Biophysical journal, 72*(3), 1335–1346.

Washizu, K. (1955). *On the variational principles of elasticity and plasticity*. Massachusetts Institute of Technology Aeroelastic and Structures Research Laboratory.

Weiss, H. (2002a). "Dynamics of geometrically nonlinear rods: II. Numerical methods and computational examples". *Nonlinear Dynamics, 30*(4), 383–415.

Weiss, H. (2002b). "Dynamics of geometrically nonlinear rods: I. Mechanical models and equations of motion". *Nonlinear Dynamics, 30*(4), 357–381.

Wempner, G. (1969). Finite elements, finite rotations and small strains of flexible shells. *International Journal of Solids and Structures, 5*(2), 117–153.

Wiggins, C. H., and Riveline, D., and Ott, A., & Goldstein, R. E. (1998). Trapping and wiggling: Elastohydrodynamics of driven microfilaments. *Biophysical journal, 74*(2), 1043–1060.

Wood, W. L. (1990). *Practical time-stepping schemes*. Oxford University Press.

Wood, W. L., Bossak, M., & Zienkiewicz, O. C. (1980). An alpha modification of Newmark's method. *International Journal for Numerical Methods in Engineering, 15*(10), 1562–1566.

Wriggers, P. (2008). *Nonlinear finite element methods*. Springer.

Yamakawa, H. (1971). *Modern theory of polymer solutions*. Harper & Row.

Yan, J., Skoko, D., & Marko, J. F. (2004). Near-field-magnetic-tweezer manipulation of single DNA molecules. *Physical Review E - Statistical, nonlinear, and soft matter physics, 70*(1), 011905.

Yang, Y., Tobias, I., and Olson, W. K. (1993). Finite element analysis of DNA supercoiling. *The Journal of Chemical Physics, 98*(2), 1673–1686.

Zhang, Q., & Marszalek, P. E. (2006). Solvent effects on the elasticity of polysaccharide molecules in disordered and ordered states by single-molecule force spectroscopy. *Polymer, 47*(7), 2526–2532.

Zhang, Z., Qi, Z., Wu, Z., and Fang, H. (2015). "A spatial Euler-Bernoulli beam element for rigid-flexible coupling dynamic analysis of flexible structures". *Shock and Vibration* 2015.

Zhang, Z., and Karniadakis, G. (2017). *Numerical methods for stochastic partial differential equations with white noise*. Springer.

Zhao, Z., & Ren, G. (2012). A quaternion-based formulation of Euler-Bernoulli beam without singularity. *Nonlinear Dynamics, 67*(3), 1825–1835.

Ziegler, H. (2013). *Principles of structural stability*. Vol. 35. Birkhäuser.

Zienkiewicz, O. C., & Taylor, R. L. (2005). *The finite element method for solid and structural mechanics*. Elsevier.

Zienkiewicz, O. C., Taylor, R. L., & Zhu, J. Z. (2005). *The finite element method: its basis and fundamentals*. Elsevier.

Zupan, D., & Saje, M. (2003). Finite-element formulation of geometrically exact three-dimensional beam theories based on interpolation of strain measures. *Computer Methods in Applied Mechanics and Engineering, 192*(49–50), 5209–5248.

Zupan, D., & Saje, M. (2006). The linearized three-dimensional beam theory of naturally curved and twisted beams: The strain vectors formulation. *Computer Methods in Applied Mechanics and Engineering, 195*(33–36), 4557–4578.

The manufacturer's authorised representative in the EU is Springer Nature Customer Service Centre GmbH, Europaplatz 3, 69115 Heidelberg, Germany. If you have any concerns regarding our products, please contact ProductSafety@springernature.com

Printed and bound by CPI Group (UK) Ltd, Croydon, CR0 4YY

30/09/2026

02225310-0001